North American Mixing Forum 2022

Held at the 2022 AIChE Annual Meeting

Phoenix, Arizona, USA
13-18 November 2022

ISBN: 978-1-7138-7893-3

Printed from e-media with permission by:

Curran Associates, Inc.
57 Morehouse Lane
Red Hook, NY 12571

Some format issues inherent in the e-media version may also appear in this print version.

Printed with permission by Curran Associates, Inc. (2023)

For permission requests, please contact AIChE
at the address below.

AIChE
120 Wall Street, FL 23
New York, NY 10005-4020

Phone: (800) 242-4363
Fax: (203) 775-5177

www.aiche.org

Additional copies of this publication are available from:

Curran Associates, Inc.
57 Morehouse Lane
Red Hook, NY 12571 USA
Phone: 845-758-0400
Fax: 845-758-2633
Email: curran@proceedings.com
Web: www.proceedings.com

TABLE OF CONTENTS

ADVANCES IN COMPUTATIONAL ANALYSES OF MIXING PROCESSES

ANALYSES OF MIXING PROCESSES IN BIOREACTORS

MIXING IN MULTIPHASE SYSTEMS

DIVISION PLENARY: NORTH AMERICAN MIXING FORUM AWARDS SESSION (INVITED TALKS)

EXPERIMENTAL INVESTIGATIONS OF MIXING PROCESSES

NOVEL, UNCONVENTIONAL, OR NON-CSTR MIXING SYSTEMS

DIVISION PLENARY: IN MEMORIAM — DR. DOUGLAS E. LENG (INVITED TALKS)

MIXING SCALE-UP AND SCALE-DOWN ISSUES IN PHARMACEUTICAL PROCESSES

134a Mixing Time Determination in a Vessel with Rotating Basket (USP Dissolution Testing Apparatus 1)

Justin Pace, Chadakarn Sirasitthichoke, Piero Armenante

The USP Apparatus 1 (rotating basket) has become a preferred dissolution testing tool to evaluate the dissolution performance of pharmaceutical oral solid dosage forms and to determine drug formulations with different drug release profiles. In general, samples are taken at a specified midway position inside the dissolution media volume during the dissolution testing process to obtain the dissolution profile. The profile represents the amount of the active drug released from its dosage form over time. Therefore, of significant importance during the dissolution process is the mixing time determination, i.e., the time needed for the drug substance released from the dosage form to become dispersed throughout the vessel and especially at the sampling location. The objectives of this work are to experimentally and computationally evaluate the blend time in the USP Apparatus 1 under different operating conditions, i.e., basket mesh size, liquid fill volume, and agitation speed. Experimentally, a non-intrusive discoloration method coupled with image processing using MATLAB was conducted. Time-dependent simulations using the Lattice Boltzmann Method combined with Large Eddy simulation were additionally performed to computationally predict mixing time. Our results show that mixing time is a strong function of the basket mesh size, agitation speed, and liquid fill volume. The experimental and predicted mixing time results were in substantial agreement and, thus, fitted with a correlated equation developed for the USP 1 system. Our study could be helpful for practitioners in selecting appropriate dissolution testing as well as operating conditions, especially when the samples are removed from the medium to obtain dissolution profiles.

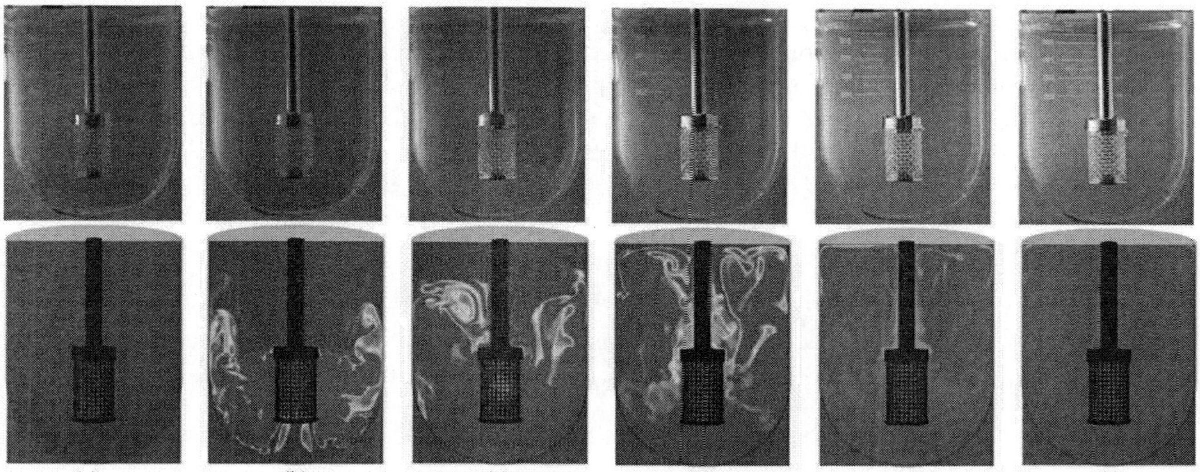

(a) (b) (c) (d) (e) (f)

134b Efficient Modeling of Polydisperse Solid-Liquid Suspensions in Stirred Tanks

Philipp Eibl, Martin Fruhwirth, Christian Witz, Johannes G. Khinast

Predicting the movement and interaction of dispersed solids with the continuous phase in stirred tanks is of key interest in a variety of industrial applications in the chemical and biopharmaceutical industry such as heterogeneous catalysis, dissolution, or microcarriers in cell culture bioreactors. Key parameters of interest are the minimum required agitation speed for off-bottom suspension (Just-suspended speed) and inhomogeneities in the particle concentration across the reactor volume in order to determine optimal feeding and sampling locations. Complexity in the prediction of this movement arises due to the simultaneous interaction of a large number of particles, making predictions based on analytical solutions of the settling behavior of single spherical particles impossible. Additional difficulty arises for non-spherical particles where the forces on the particles depend on the orientation of the particles within the flow field. While classical RANS-based CFD modeling approaches proved to be sufficiently accurate for the prediction of just-suspended speeds, transient effects, and local inhomogeneities may not be captured sufficiently accurately [1]. In this work, we use the Lattice Boltzmann Method to predict the transient three-dimensional flow field inside stirred tanks and employ an Euler-Lagrange approach via the discrete particle method (DPM) for modeling the dispersed solid phase. This approach was shown to be suitable for small particle diameters below the numerical grid size. In order to decouple the size dependency of the dispersed particles from the numerical grid, several models are reported in the literature [2]. Here we present a coupling approach to generalize the Lagrangian dispersed phase and resolve this grid dependency based on the two-grid method. The simulation runtimes are shown to be suitable for application in the aforementioned industrial use cases by integrating and running the presented model on the commercial GPU-based modeling software SIMVANTAGE (simvantage.com).

[1] Delafosse et al. "Solid-liquid suspension of microcarriers in stirred tank bioreactor – Experimental und numerical analysis", Chem. Eng. Sci 180, 2018, p52-63

[2] Zhang et al. "Grid-indipendent Eulerian-Lagrangian approaches for simulations of solid fuel particles combustion", Chemical Engineering Journal 387, 2020

134c Optimization of a Full-Scale Disinfection Unit Using CFD at Pero WWTP in Milan

Cesare Piacezzi, Giacomo Bellandi, Alejandro Claro Barreto, Roberta Muoio, Roberto Di Cosmo, Davide Scaglione, Usman Rehman, Ingmar Nopens

Introduction

In view of coping with more stringent regulatory standards for secondary effluent disinfection, Gruppo CAP is using CFD (Computational Fluid Dynamics) to optimize the 2 disinfection units of one of their biggest wastewater treatment plants (WWTPs). The WWTP of Pero treats 30% of the municipal wastewater of Milan's metropolitan area (Italy), and therefore it is of extreme importance for the surface water quality. The aim of the study was to initially assess and then optimize the disinfectant dosage in the last treatment step of the Pero's WWTP before discharging into the receiving water body. In this work, CFD was used to identify the most effective location(s) and strategy to cope with the operational variability of the WWTP and the regulatory guidelines. Within the collaboration with Gruppo CAP and AM-Team, CFD modelling was used to perform several virtual experiments at different operational scenarios avoiding onsite tests (i.e., risky, expensive, and time consuming). Some of the scenarios analysed in this project are reported in this abstract. The results highlighted the importance of the location of both the dosage and the mixing device, especially when updating an existing installation to new regulatory requirements (Pero WWTP E. Coli limit value changed form 5000 UFC/100 ml to 1000 UFC/100 ml in 2021 and it will be 500 UFC/100ml in 2022).

Methodological Approach

The geometry of the entire disinfection unit (Figure 1) was derived from the database of Gruppo CAP, and digitally recreated to obtain a 3D domain. There are 2 influent streams to the disinfection unit: the main one comes from an inlet channel (on the top of Figure 1), after the biological treatment. In case of heavy rain events, part of the influent of the WWTP bypasses the biological treatment, and enters the disinfection unit through the bypass channel. The disinfection unit is composed of 2 reaction serpentines; each of them consists of 2 parallel channels that run straight for about 65 m, before taking a 180° U-turn.

The CFD simulations were performed in ANSYS Fluent v21R1 in steady state. The flow and disinfectant transport were simulated by means of a RANS (Reynolds-Averaged Navier-Stokes) approach, specifically with a Realizable k-ε turbulence model, and standard wall functions (Versteeg and Malalasekera, 2007). The mixers were simulated with a Multiple Reference Frame approach (ANSYS, Inc. (2021)). The VOF multiphase model was used to calculate the water level in the plant.

Results and Discussion

In the normal plant operation, the influent coming from the main channel only enters the 1st disinfection serpentine (the gate that connects the 1st and 2nd serpentine is closed), as shown in Figure 2. In the initial configuration the disinfectant dosage point is located on the corner of the basin before entering the serpentine, as shown in the digital recreation of the geometry (Figure 3). For this scenario, an average flowrate of 5000 m3/h (normal conditions) was considered. The results of the simulation of this scenario are shown in Figure 4. The water coming from the inlet channel is accelerated by the gate that leads to the basin with the mixer, before entering the 2 channels of the serpentine. Note that the dosage point is located on a stagnation zone on the side of the basin (the mixer is not able to create full mixing in the basin). This negatively affects the disinfectant distribution. Referring to Figure 4, the variable used to assess the disinfectant distribution relates the local disinfectant fraction to the desired one (if >1, there is excess of disinfectant, while if <1 there is lack of disinfectant). As a consequence of the flow pattern, the disinfectant accumulates on the side of the basin, and mainly enters only one of the two channels (Channel 1). Specifically:

- Channel 1: Local fraction / Desired fraction = 1.8
- Channel 2: Local fraction / Desired fraction = 0.14

To conclude, with this configuration the dosing strategy is not optimal.

Therefore, an improved configuration was suggested and tested, shown in Figure 5. The dosing point was moved in the middle of the inlet channel, with a mixer placed afterwards. This scenario was tested with a high influent flowrate (9000 m3/h). The results of this scenario are shown in Figure 6. The disinfectant plume is broken by the action of the mixer placed after the dosage point. As a result, a good disinfectant distribution is observed even before entering the basin that leads to the serpentine. Therefore, the new configuration proved to bring a substantial improvement in terms of disinfectant usage, as it is well mixed and equally distributed in the channels of the serpentine. However, a potential issue was detected by analysing the hydraulics: the water flowrate coming from the inlet channel is not equally split between the two channels of the serpentine. Despite the action of the impeller in the basin, the path from the turn to Channel 1 is in fact more favourable. As a result, at 9000 m3/h, 70% of the total flowrate enters Channel 1. Such splitting inequality increases as the influent flowrate increases (at 5000 m3/h the splitting was close to 50% - 50%). As a conclusion, at high influent flowrates, the HRT in the serpentine might not be sufficient, because of the high velocity in one of the two channels.

The effect of both disinfection lines (Serpentine 1 and 2) operating together was analysed in the following scenario. In such plant configuration, both the influent coming from the main channel and the bypass stream enter the disinfection unit. The geometry used for this simulation is shown in Figure 7. All gates connecting the serpentines are open, so there are no constraints for the flow. The weirs in the end of both serpentines are included in the model, as they are crucial to determine the water level. For practical reasons, the whole serpentine lengths were not modelled; the weirs were instead placed in the 1st leg of the serpentines. This scenario was modelled under the following conditions: 7000 m3/h as the influent flowrate from the main channel, and 1000 m3/h from the bypass channel. The results are shown in Figure 8. Thanks to on-site measurements, a 28 cm difference in the weirs height between the two serpentines was detected (the weir at the end of Serpentine 2 is 28 cm lower than the one of Serpentine 1). As a consequence, the path through Serpentine 2 is favoured. As shown in Figure 8, the difference in the weir height is such that the total flowrate (8000 m3/h) only enters Serpentine 2. In other words, the overflow height over the weir of Serpentine 2 is lower than the height of the weir of Serpentine 1. As a consequence, stagnation is observed in Serpentine 1. To conclude, if both disinfection lines are operating together, a whole reaction serpentine would be unused.

Conclusions

A full-scale disinfection unit was analysed by CFD modelling. The study allowed to analyse the current state of the plant and, based on that, to spot opportunities to improve it. Specifically, with the original dosage point in the first disinfection line, the disinfectant distribution in the reaction serpentine is not homogeneous, and so the disinfectant is not being optimally used. Thanks to this study, a new dosage location was identified, allowing to save a significant amount of disinfectant (≈7 times less). It was also discovered that in the first disinfection line, at high flowrates, the splitting of the total flowrate into the serpentine channels is not equal; this might lead to HRT issues. Also, when both disinfection lines are operating together, the whole influent flow only enters Serpentine 2. This work helped Gruppo CAP to optimally use the disinfectant and to reduce operational costs.

References

ANSYS, Inc. (2021) *ANSYS Fluent Theory Guide*

Versteeg, H. K. and Malalasekera, W. (2007) *An Introduction to Computational Fluid Dynamics: The Finite Volume Method.* Harlow, England: Pearson Education.

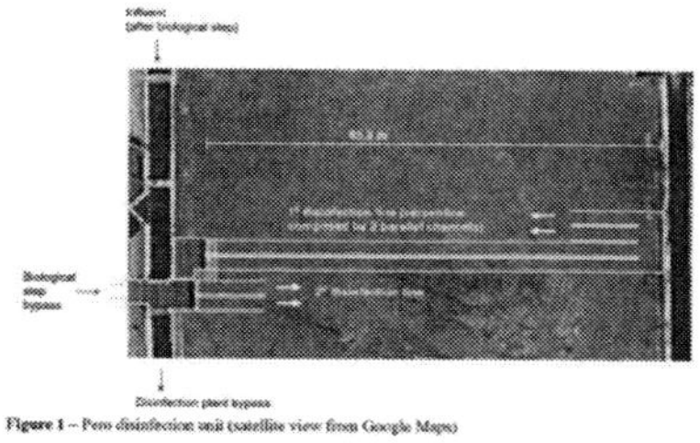

Figure 1 – Pero disinfection unit (satellite view from Google Maps)

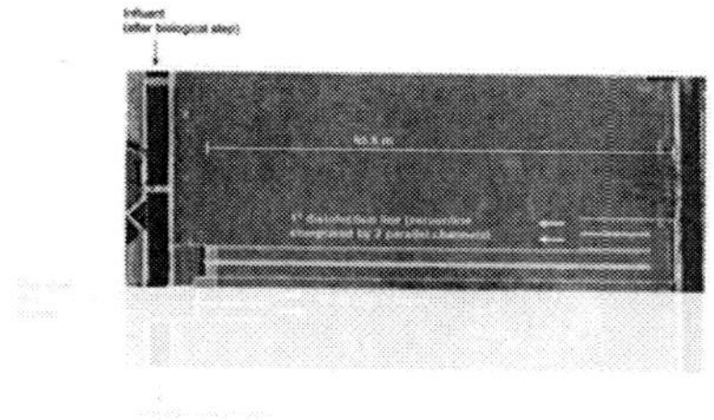

Figure 2 – Influent from the main channel, into the 1st disinfection serpentine.

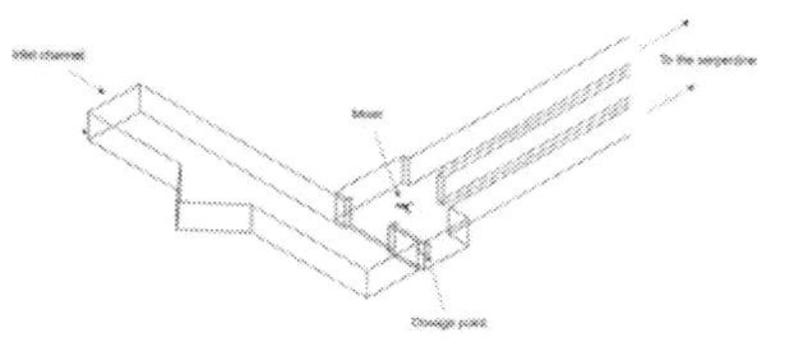

Figure 3 – 1st scenario: influent from the main channel, into the 1st disinfection serpentine. Dosing point on the corner of the basin.

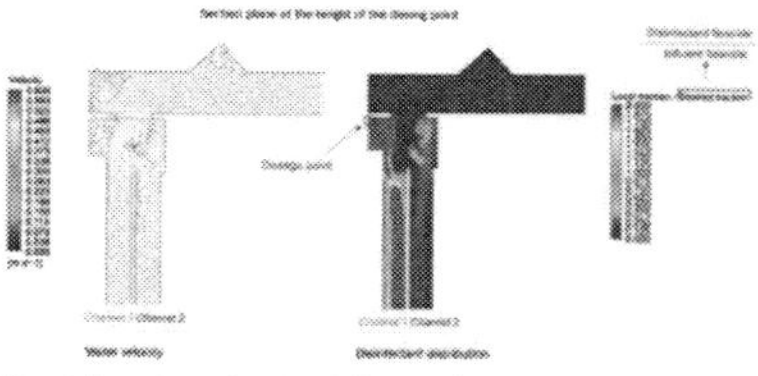

Figure 4 – 1st scenario: water flow pattern and disinfectant distribution

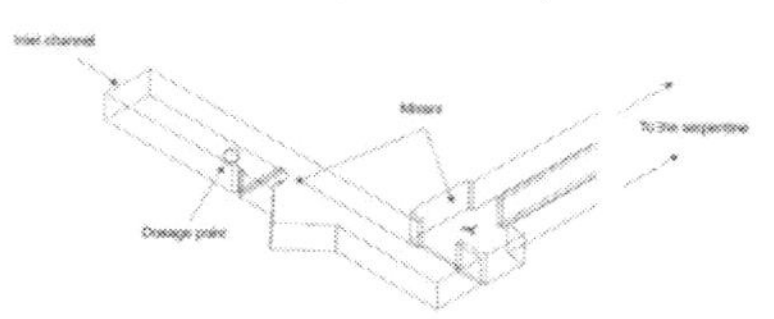

Figure 5 – First plant configuration, with changes: influent from the main channel, into the 1st disinfection serpentine. Dosing point in the inlet channel, with a mixer placed afterwards.

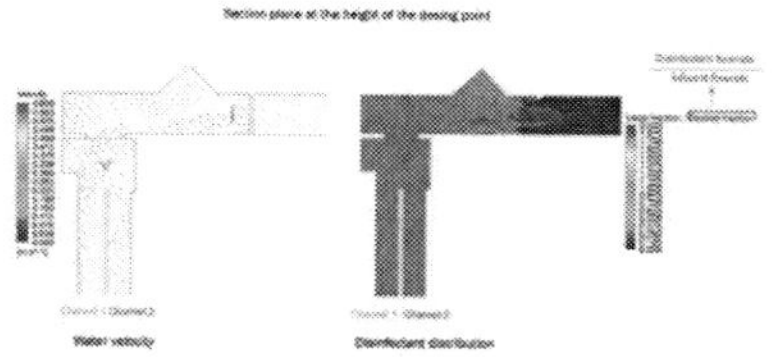

Figure 6 – 2nd scenario: water flow pattern and disinfectant distribution

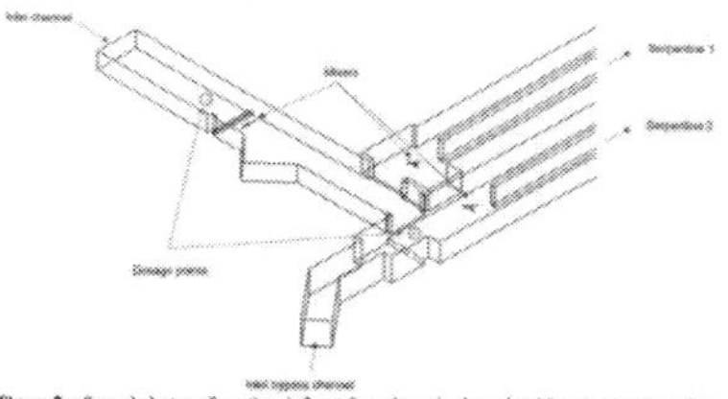

Figure 7 – Second plant configuration: influent from the main channel and bypass, two serpentines communicating.

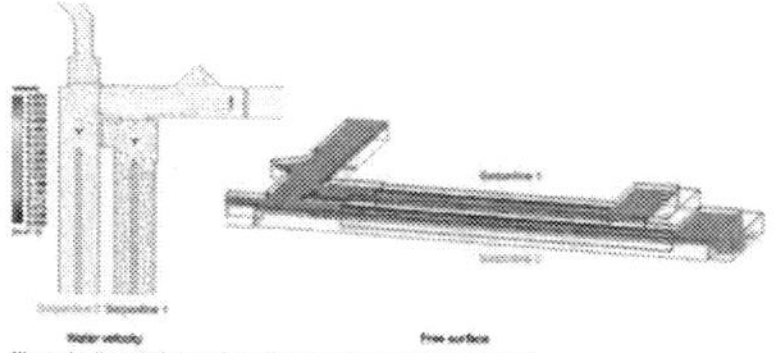

Figure 8 – Second plant configuration: water flow pattern and free surface

134d Lattice-Boltzmann Computational Fluid Dynamics (CFD) Simulation of Jet Mixing in Tanks

John Thomas, Michael Poirier

Jet mixers have been used for a number of years in process industries. Common applications include adding acid or base for pH adjustment, ensuring the tank is well mixed prior to sampling for chemical analysis, and adding material to adjust the composition of the tank contents. In addition, they are widely used at U.S. Department of Energy sites to mix and process waste in ~ 1 million gallon tanks. Because of the size of the vessels and the radioactivity of the material in these tanks, testing at full-scale or pilot-scale is problematic and expensive. An alternative to assist in the design and operation of jet mixers in these large tanks is to use computational fluid dynamics (CFD). The authors will discuss the development of CFD models using the M-Star® lattice-Boltzmann software to simulate jet mixing in tanks.

The authors developed models of mixing in tanks using stationary jet mixers with a single jet, jet mixers with two opposing jets, and rotating jet mixers with two opposing jets. The models included the inlet or pump suction and the recirculation of the inlet stream to the jet nozzle. A refinement zone was added to the model to allow for a small lattice spacing near the jet nozzle discharge, to capture the physics of the jet formed, and a coarser lattice spacing away for the jet nozzle, to reduce the computation time. The jet nozzle diameter and nozzle discharge velocity were varied to challenge the models.

The results of the simulations performed with these models were compared with experimental data in the technical literature. For the stationary jet mixers, the comparisons were of miscible liquid blend times in vessel of diameters between 0.6 and 36 meters. The rotating jet mixer model was used to simulate the mixing of Bingham plastic fluids in a tank. The comparisons were of the effective cleaning radius of the jet mixing, which is a measure of its zone of influence. The effective cleaning radius is analogous to the mixing cavern in an impeller mixed tank. The results of the simulations and the comparisons to experimental data will be discussed.

134e Accelerating Biopharmaceutical Mixing Process Development with Hybrid Modeling and Computational Fluid Dynamics

Ross Kenyon

Mixing operations in the biopharmaceutical manufacturing process, though straightforward in nature, still require extensive amounts of experimentation and validation; both aspects being resource and time intensive. Mixing is a highly dimensional problem as its key input parameters - impeller speed, mixing time, batch volume, product viscosity and vessel geometry - impact multiple aspects of product quality. A solid understanding of these relationships is critical to ensure patient safety and regulatory acceptance. Computational techniques have opened the door to significantly reducing the amount of experimentation and accelerating qualification efforts for new processes. The latter has become even more important in today's world, with a greater need to respond rapidly to new public health threats such as COVID-19. We will present a case study of how Computational Fluid Dynamics (CFD) was recently used to accelerate the mixing process development timeline for Regeneron's REGN-COV2 antibody cocktail by mapping mixer performance and mitigating risks to product quality. We will also detail a framework of how high-throughput experiments, online conductivity measurements, data-driven models and mechanistic models can all be combined to further enhance process development. Finally, we will share a verification and validation effort - aligned with the ASME's VVUQ 40 standard and the FDA's reporting guidance for computational modeling - that will serve to increase model credibility and expand the role of models in decision making. Putting each of these elements together, we envision the lean process development roadmap of the future: where computational methods supplant physical testing, product-specific development packages are reduced to a minimum, and rapid process development is a reality.

FLOW SIMULATION OF BIOREACTORS USING ENTROPIC LATTICE BOLTZMANN METHOD

Sai Nithin Reddy Adidela[1], *Chakradhar Thantanapally*[1*],
Vinay Kariwala[1], Salvatore Arcidiacono[1], Santosh Ansumali[12]
[1]SankhyaSutra Labs, Bangalore India,
[2]Jawaharlal Nehru Centre For Advanced Scientific Research
*chakradhar@sankhyasutralabs.com

Abstract

In the biopharmaceutical industry, stirred-tank reactors are extensively used due to their mixing performance, efficient mass, and heat transfer. Characterization of stirred-tank reactor and proper understanding of the mixing process is essential for streamlining the production of biopharmaceuticals. Stirred-tank reactors are typically characterized by their power consumed, which is accurately predicted by various steady-state Reynolds-averaged Navier–Stokes (RANS) simulations. In order to study the inherently turbulent, multi-phase, non-stationary flow characteristics of the mixing process, and to accurately predict the mixing time, efficient transient schemes are needed. Direct Numerical Simulation (DNS) of such a complicated setup is compute intensive. In literature, these complexities are dealt by a number of modelling techniques. The objective of the current work is to simulate flow in single-phase stirred-reactor comprising of a single and double Rushton impellers respectively by employing the Entropic Lattice Boltzmann Method (ELBM) without any explicit modelling.

Introduction

Continuous stirred-tank reactor, also known as mixing-tank reactor or stirred-vessel is a specific type of bioreactor in which reactants and reagents are mixed with the help of an agitator[1]. These are widely used in biopharmaceutical industry due to their mixing performance, efficient mass, and heat transfer. Experimental results are available for single-phase as well as multi-phase stirred-tank reactors in literature. Several advanced Lagrangian Particle Tracking techniques like 4D-PTV made it possible to visualize the three dimensional flow structures and provide insights into the mixing process and also serves as a base for numerical simulations[2]. However, integrating experimentation into the design cycle of biopharmaceutical equipment is not feasible owing to the material cost and complexity involved in conducting experiments on scaled-down models.

Computational Fluid Dynamics (CFD) is becoming increasingly popular in the biopharmaceutical industry[3] aiding or replacing experiments. CFD plays a crucial role in development of scaled-down laboratory models, studying the underlying physical and chemical process, optimizing design[4], rapid prototyping and production of scale-up models[5]. By doing so, CFD significantly decreases the design-cycle of biopharmaceutical equipment thus overall resulting in a reduced time to market in an extremely competitive industry[2].

Steady-state Reynolds-averaged Navier-Stokes (RANS) simulations for stirred-tank reactors are extensively presented in literature. Moving bodies are modelled using techniques like steady-state Multiple Reference Frame methods[6] to account for impeller rotation. RANS approach with steady-state MRF technique was very capable in predicting the steady state characteristics such as power number which is typically used to characterize stirred tank reactors[6].

In order to completely understand the physics of the mixing process, in addition to predicting the power number, it is very important to resolve transient flow structures as the flow inside a stirred-tank reactor is inherently turbulent and non-stationary[2]. Because of the complexity of the set-up involving moving bodies, it becomes very difficult to obtain meaningful results with affordable resolutions using standard DNS techniques. Hence, Large-eddy Simulations (LES) where moving bodies are modelled using modelling techniques such as Sliding Mesh technique[7] are becoming the industry standard for mixing time simulations.

In the current study,in-house solver SankhyaSutra Taral© implementing the Entropic Lattice Boltzmann Method (ELBM) is employed to simulate a single-phase stirred-tank reactors with a single and double Rushton impellers respectively. The solver makes use of D3Q27 lattice model and a structured Body Centered Cubic (BCC) arrangement of grid points for spatial discretization. No explicit turbulence models have been used in this study. The solver is capable of meshing multiple complex moving geometries. The geometry and the method employed are discussed in the further sections.

Numerical Methodology

Reactor Setup with single Rushton impeller

Detailed geometry of the stirred-tank reactor can be found in Fig. 1 of [8]. A geometry based on the description given in the reference can be seen in Fig. 1

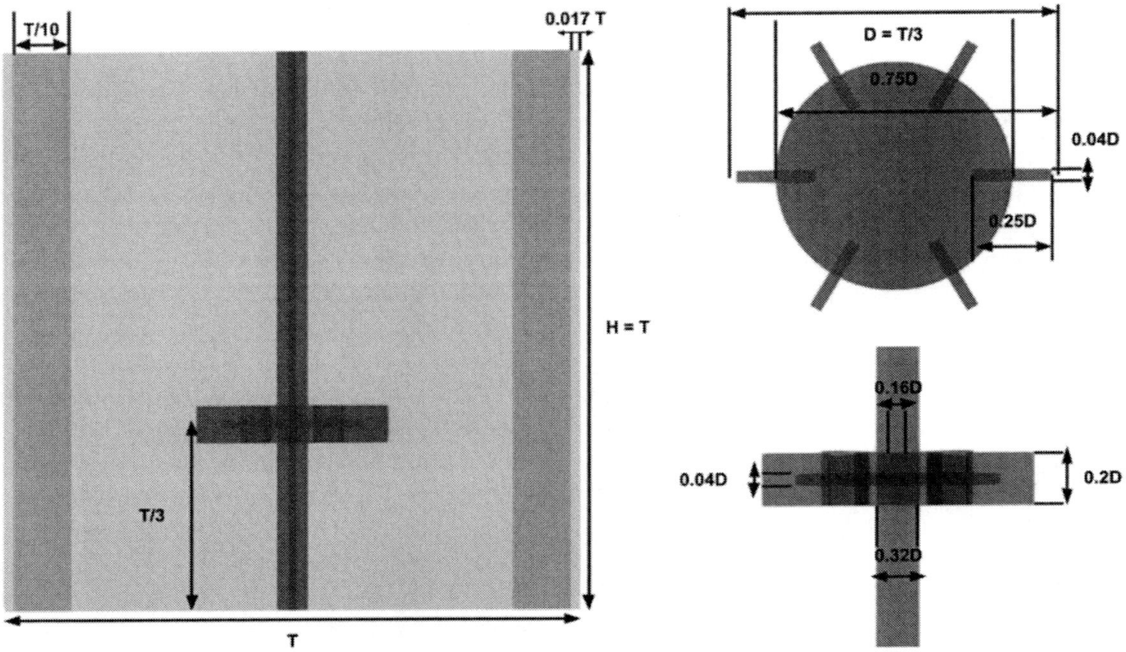

Figure 1: Geometry of the stirred tank reactor with single Rushton impeller

The setup consists of a cylindrical tank with diameter $T = 150mm$ and a fluid-filled height of $H = T = 150mm$ from the base of the tank. The tank consists of 4 baffles of width $H = 150mm$, thickness $0.1T = 15mm$, arranged uniformly around its outer walls with a baffle clearance of $0.017T = 2.55mm$. The tank consists of one Rushton impeller of diameter $D = \frac{T}{3} = 50mm$. The clearance from the bottom of the tank to the center of the impeller is $C = \frac{T}{3} = 50mm$. The central-disk of the impeller has an outer diameter of $0.75D = 37.5mm$, an inner diameter of $0.32D = 16mm$ and a height of $0.04D = 2mm$. The hub connecting the impeller and shaft has an outer diameter $0.32D = 16mm$, an inner diameter of $0.16D = 8mm$ and a height of $0.2D = 10mm$. The Rushton impeller has 6 blades and each blade has a length of $0.25D = 12.5mm$, a height of $0.2D = 10mm$, and a thickness of $0.04D = 2mm$. The shaft has a diameter of $0.16D = 8mm$ and a height of $H = 150mm$.

Reactor setup with double Rushton impellers

Detailed geometry of the multi-phase bioreactor is found in Fig. 1 of [2]. A geometry based on the description given in the reference is prepared with FreeCAD as shown in Fig. 2

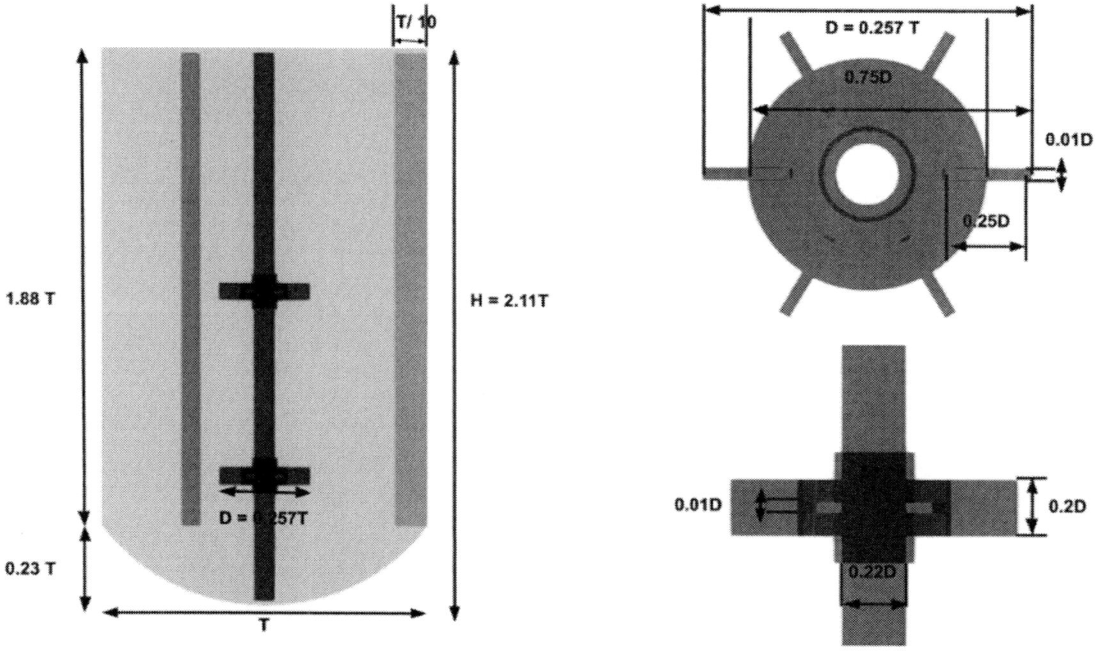

Figure 2: Geometry of the stirred tank reactor with 2 rushton impellers

The geometry consists of a cylindrical tank with diameter $D = 140mm$ and fluid-filled height $H = 228mm$ from the base of the tank. The tank has a hemi-spherical end at the bottom. Volume of the fluid-filled tank is 2.8 liters. The setup consists of two rushton impellers of diameter $d = 36mm$. The hub connecting the impeller and shaft has an outer diameter $10mm$ and and an inner diameter equal to the diameter of the shaft $8mm$. The central-disk of the impeller has a diameter of $0.75d = 27mm$. The rushton impeller has 6 blades and each blade has a length of $0.25d = 9mm$, a height of $0.2d = 7.2mm$, and a thickness of $0.01D = 1.3mm$

Entropic Lattice Boltzmann Method

Lattice-Boltzmann methods originating from Lattice Gas Automata are becoming increasingly popular in Computational Fluid Dynamics domain given their advantages in dealing with complex boundaries, parellelization of the algorithm [9] [10]. In contrast to traditional CFD methods which solve the conservative equations of macroscopic properties such as mass, momentum, and energy, LBM models the fluid as density on ficticious particles performing collision and advection steps consecutively over a lattice grid. The fluid density $\rho(\vec{x}, t)$ at position x and time t on the lattice designated by $DnQm$ where Dn stands for n-Dimensions and Qm stands for m-speeds, has a number of components $f_i(\vec{x}, t)$, $0 \leq i \leq m$. These individual components are evolved in time by the following steps.

Collision step

The Bhatnagar Gross and Krook (BGK) collision model is given by

$$f_i(\vec{x}, t + \delta t) = f_i(\vec{x}, t) + \frac{f_i^{eq}(\vec{x}, t) - f_i(\vec{x}, t)}{\tau_f}$$

where $f_i^{eq}(\vec{x}, t)$ is the i_{th} component of equilibrium density anologous to Maxwell's distribution, based on the current, local fluid density. τ_f is the characteristic time-scale over which the local density relaxes to equilibrium. τ_f is a measure of kinematic viscosity.

Advection step

$$f_i(\vec{x} + c_i, t + \delta t) = f_i(\vec{x}, t)$$

where c_i is the i_{th} componennt of lattice velocity.

The macroscopic properties of the fluid at any instance are calculated by

$$\rho = \sum_{i=0}^{m} f_i^{eq}$$

$$\rho\vec{u} = \sum_{i=0}^{m} f_i^{eq}\ \vec{c_i}$$

Entropic modelling

To address the instabilities of traditional Lattice Boltzmann Methods for under-resolved simulations [11, 12], Entropic Lattice Boltzmann Method was developed [13, 14, 15, 16, 17]. The ELBGK model is unconditionally stable and is instrumental in performing accurate numerical simulations of high-Re, turbulent flows without any explicit modelling, respecting the Second Law of Thermodynamics. Instead of using standard Simple Cubic (SC) lattice for spacially discretizing the computational domain, a Body-Centred Cubic (BCC) lattice was used to further improve the capability of ELBM making use of recently introduced crystallographic LBM [18].

Numerical methodology

Model and Grid (Reactor setup with single Rushton impeller)

D3Q27 Lattice model was used with a computational domain for the simulations is $-75mm \leq X \leq 75mm$, $-75mm \leq Y \leq 75mm$ and $0mm \leq Z \leq 150mm$ A no-slip boundary condition, Diffuse bounce-back[19] is applied on tank walls, top and bottom. The moving walls of the impeller and shaft are tracked using the level-set method[20]. A modified no-slip boundary condition for moving walls, Diffuse bounce-back[19] where the bounced back populations are equilibriated using the moving wall velocity is applied on impeller walls and shaft walls. The impeller rotated with an angular velocity of 2672 rpm. Considering the impeller-tip velocity as the characteristic velocity and impeller diameter as the characteristic length scale, the Reynold's number $\frac{V_{tip}D}{\nu}$ is 23000. The Mach number is 0.1.

Model and Grid (Reactor setup with double Rushton impellers)

D3Q27 Lattice model was used with a computational domain for the simulations is $-70mm \leq X \leq 70mm$, $0 \leq Y \leq 228mm$ and $-70mm \leq Z \leq 70mm$ A free-slip boundary condition is applied at the top free-surface of the liquid. A no-slip boundary condition, Diffuse bounce-back[19] is applied on tank walls. on impeller walls and shaft walls. The impellers are rotated with an angular velocity of 250 rpm. Considering the impeller-tip velocity as the characteristic velocity and impeller diameter as the characteristic length scale, the Reynold's number $\frac{V_{tip}D}{\nu}$ is 17000. The Mach number is 0.1.

Results

Reactor with single Rushton impeller

The flow properties are captured at equal intervels of 20 times per rotation of the impeller in 4 vertical planes, each between a pair of baffles. The results presented are an azimuthal average of the 4 vertical planes, averaged over 15 rotations of the impeller.

Two large circulation loops are seen as dominant features of a radial flow impeller [8]. One loop is seen below the impeller going downward near the tank wall and upward near the shaft. The other loop is seen above the impeller going upward near the tank walls and downward near the shaft. The upper circulation loop does not extent to the top of the tank indicating that the portion after this separation point, the fluid is badly mixed. This separation point is predicted to be at $\frac{z}{T} = 0.8$ by experiments, RANS approach and $\frac{z}{T} = 0.73$ using the LES approach. SSLabs D3Q27BCC with the same resolution used in RANS and LES i.e., 0.675mm corresponding to 80 parts per impeller diameter predicts the separation point at $\frac{z}{T} = 0.73$ as shown in fig 3

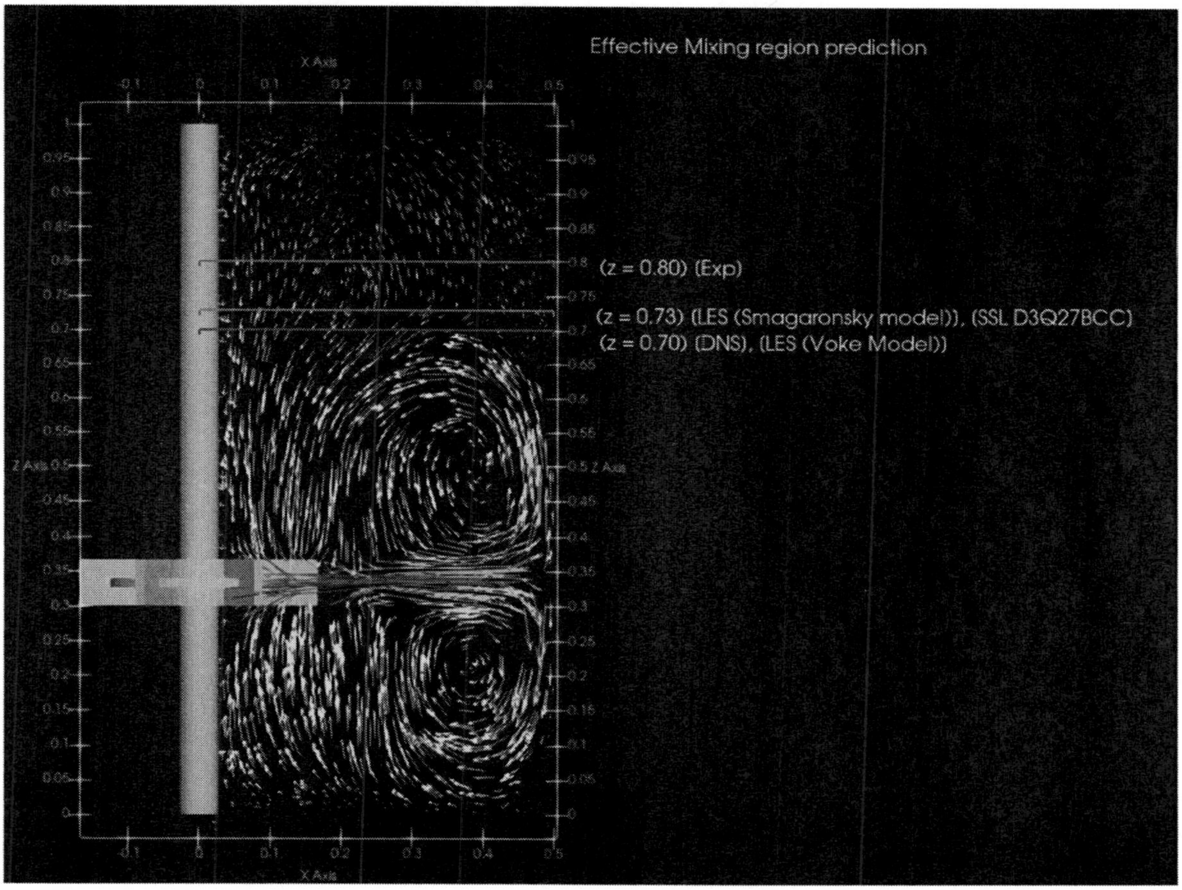

Figure 3: Comparision of the separation point of the Upper circulation loop

The shape and center of the lower circulation loop is in agreement between experimental, RANS and LES approaches, predicting the center to be at $\frac{x}{D} = 0.38 and \frac{z}{D} = 0.22$. SSLabs D3Q27BCC with the same resolution used in RANS and LES i.e., 0.675mm corresponding to 80 parts per impeller diameter predicts the center of the lower circulation loop at $\frac{x}{T} = 0.73 \frac{z}{T} = 0.20$ as shown in fig 4

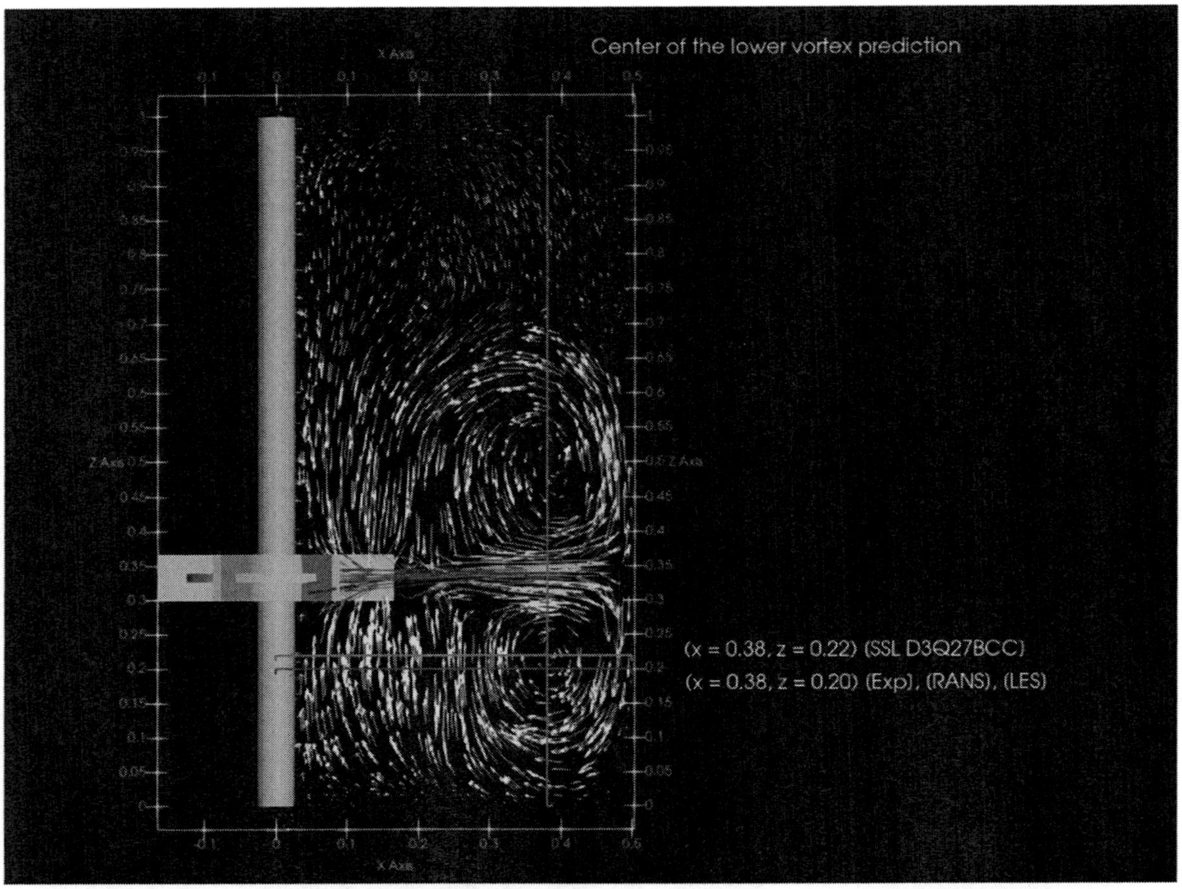

Figure 4: Comparision of the center of the Lower circulation loop

First order quantities of Radial, Tangential velocities and second order quantity of Turbulent kinetic energy are plotted along the height of the tank at various radial positions of $\frac{r}{T} = 0.183, 0.25, 0.317$ as seen in fig 5

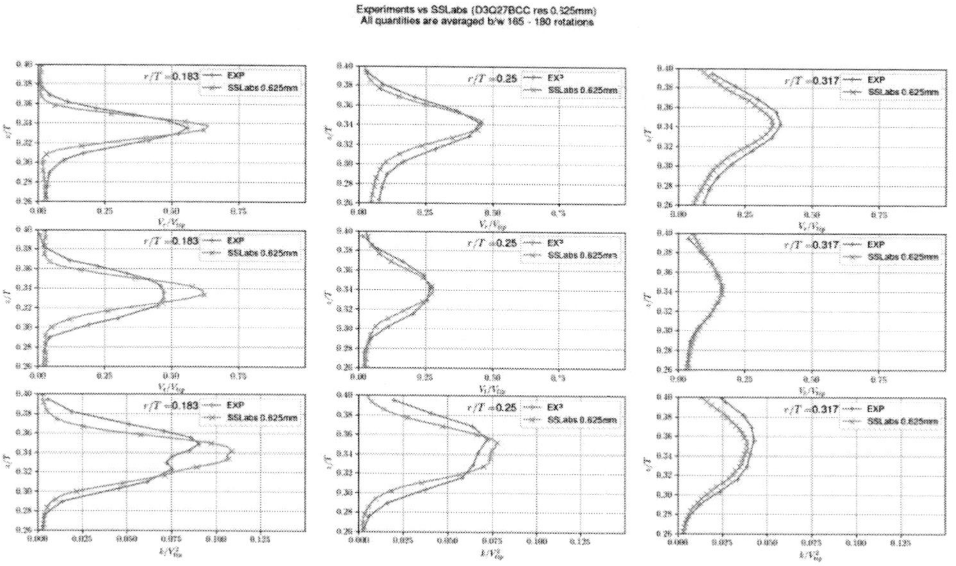

Figure 5: Averaged Velocity and TKE Profiles along height at various radial positions

A grid convergance study is run with resolutions corresponding to 40, 80, 120 parts per impeller diameter and it can be seen that the first order as well as second order quantities are in a better agreement with the experimental results at PPD 120 as shown in fig 6.

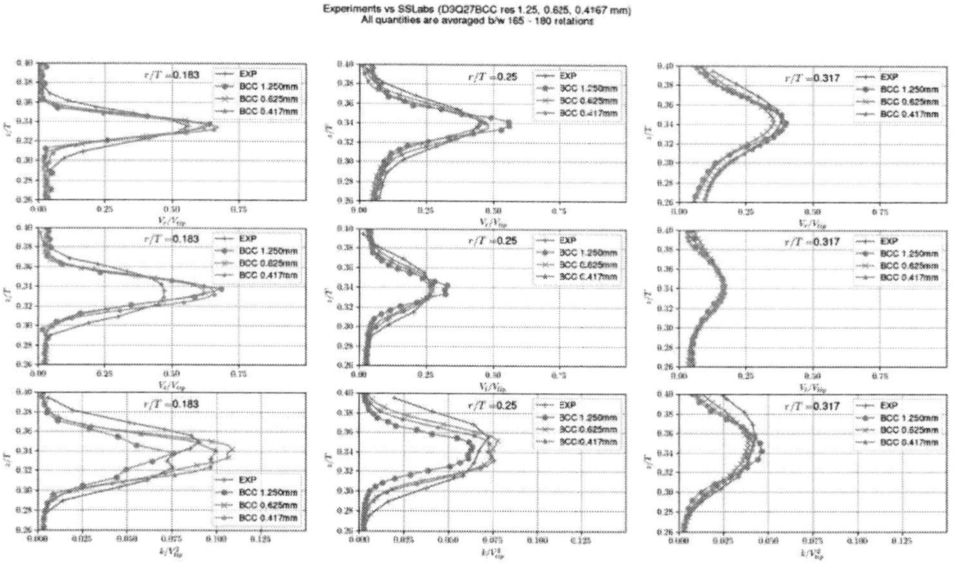

Figure 6: Grid convergence study

A comparision of SSLabs D3Q27BCC, RANS, LES with the same resolution corresponding to 80 parts per impeller diameter can be seen in fig 7

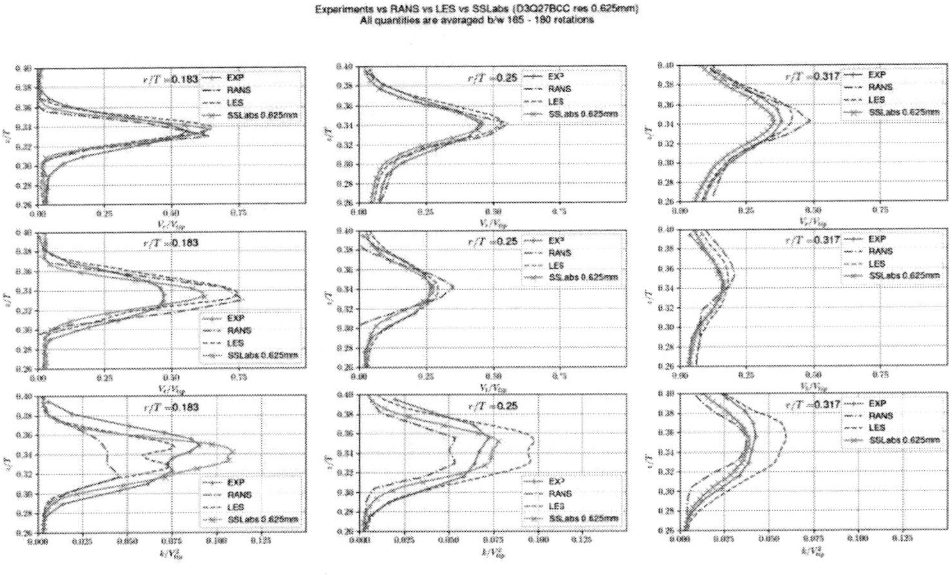

Figure 7: Comparision with RANS and LES approaches

A comparision of SSLabs D3Q27BCC, D3Q27SC with the same resolution corresponding to 80 parts per impeller diameter can be seen in fig 8

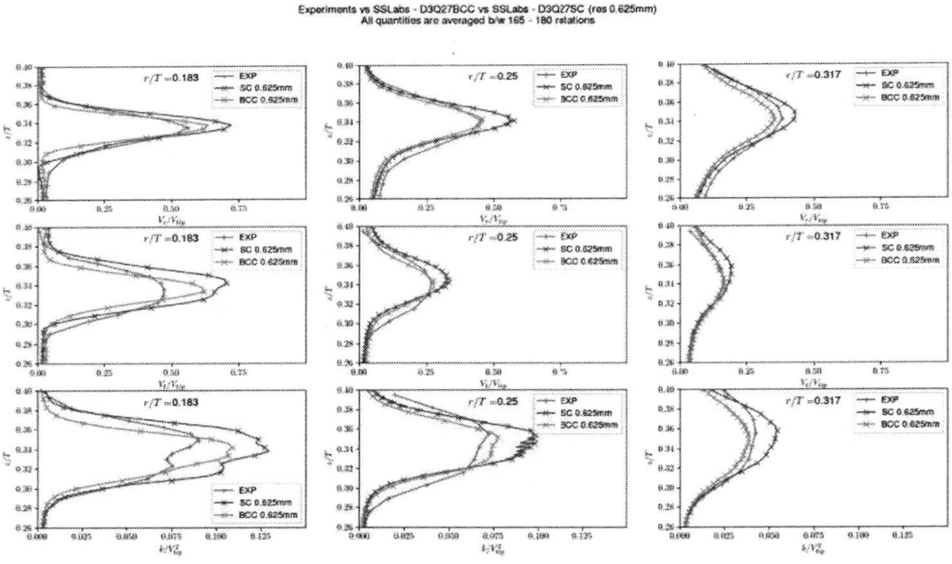

Figure 8: Comparision of SC and BCC grids

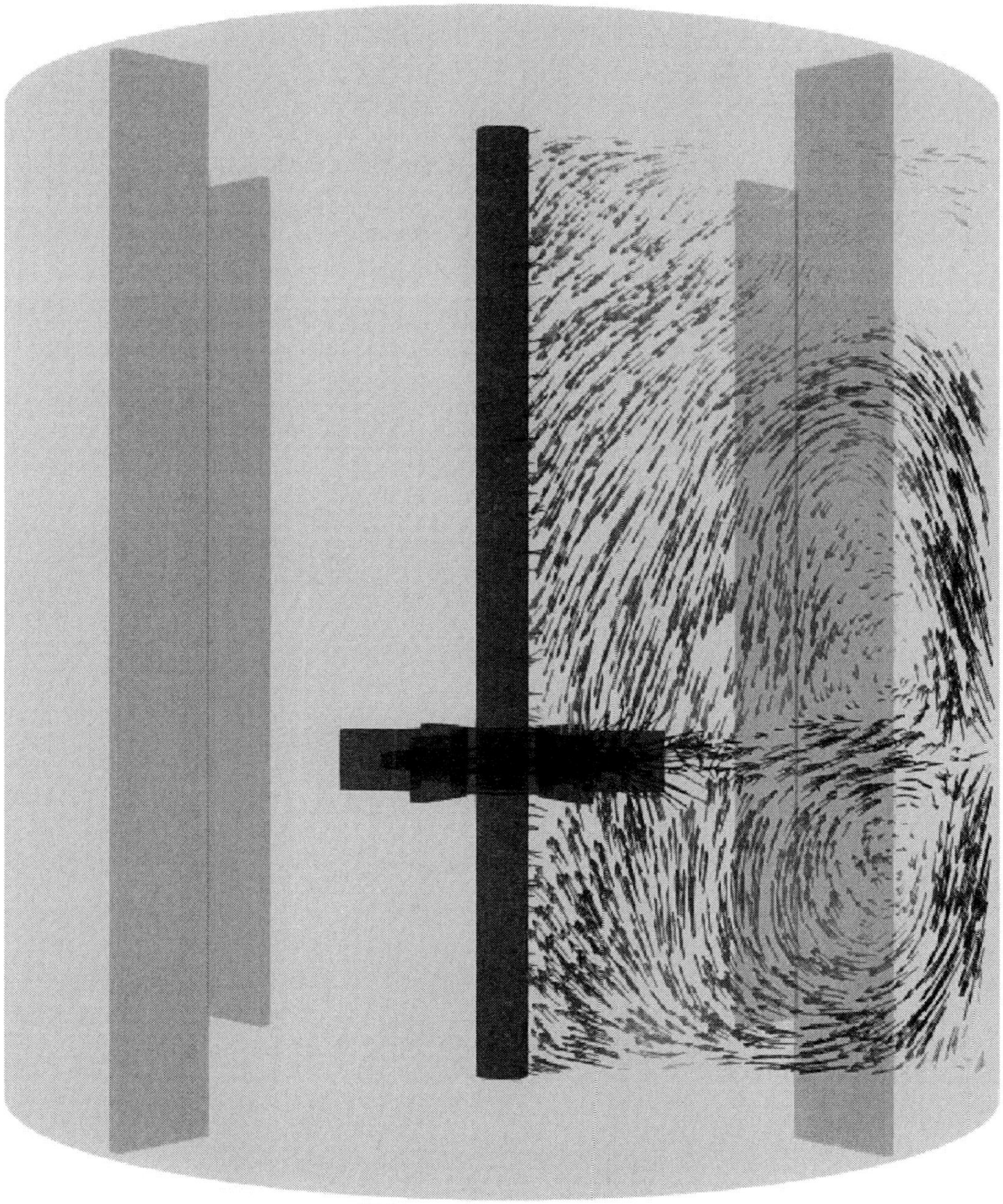

Figure 9: Visualization of the circulation loops using velocity magnitude glyphs

Reactor with double Rushton impellers

Time and azimuthally averaged velocity profile in the fluid stream of the top and bottom impellers are seen in fig 10 and fig 11 respectively. Midplanes of the impellers are considered for averaging. It is to be noted that experimental reference data is not available for this plot.

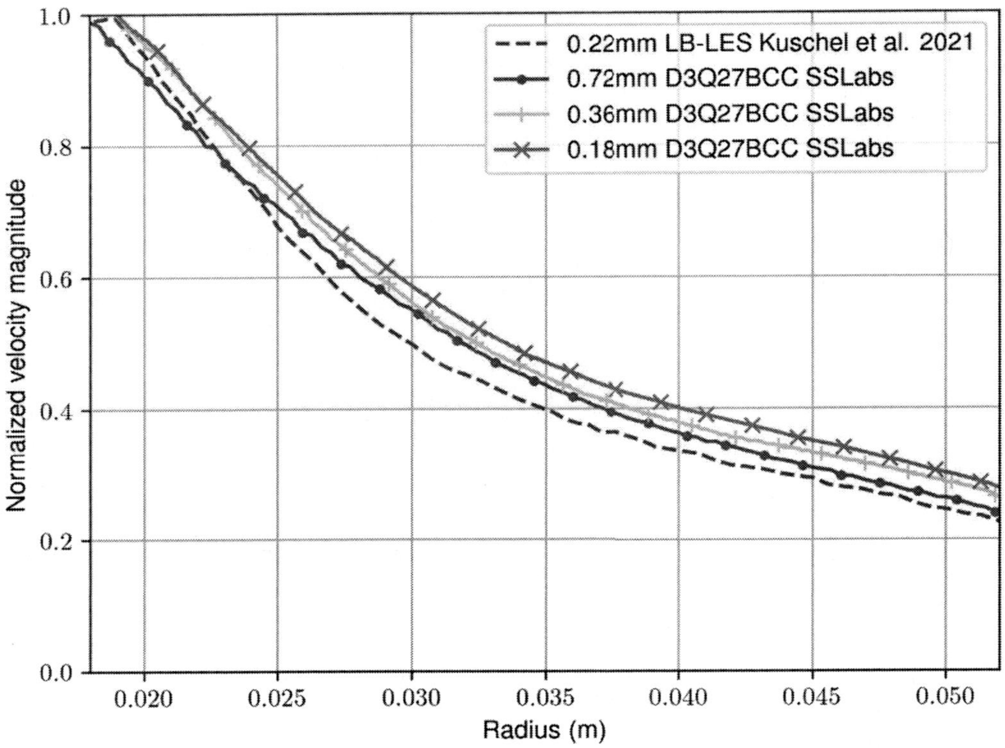

Figure 10: Time and azimuthally averaged profiles of normalized velocity magnitude in top impeller fluid stream

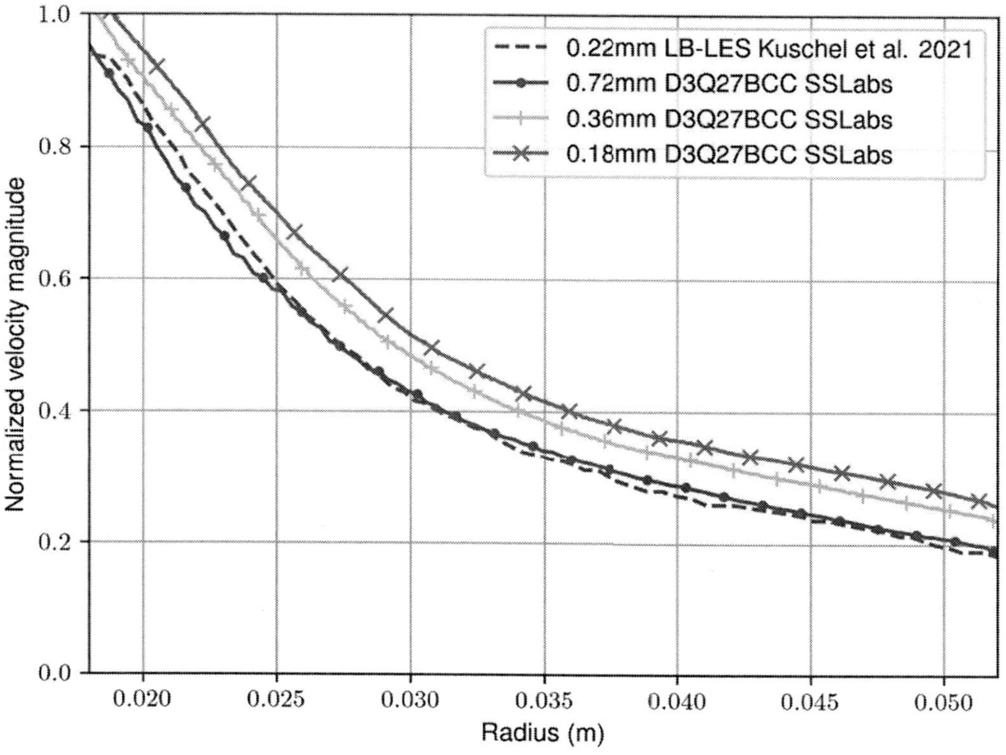

Figure 11: Time and azimuthally averaged profiles of normalized velocity magnitude in bottom impeller fluid stream

Time and azimuthally averaged velocity profiles at different radial positions in the tank are shown in fig 12.

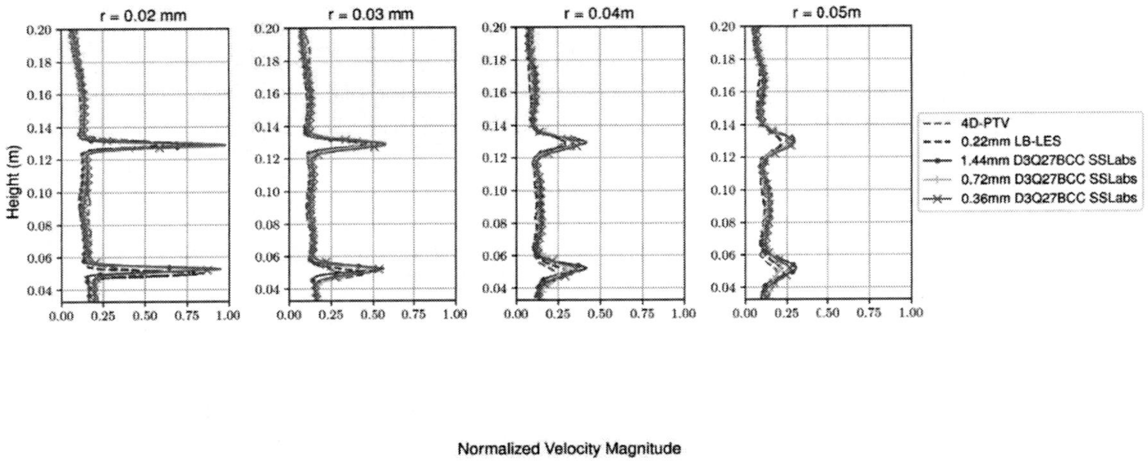

Figure 12: Comparison of the dimensionless axial time- and azimuthal-averaged velocity profiles at different radial positions

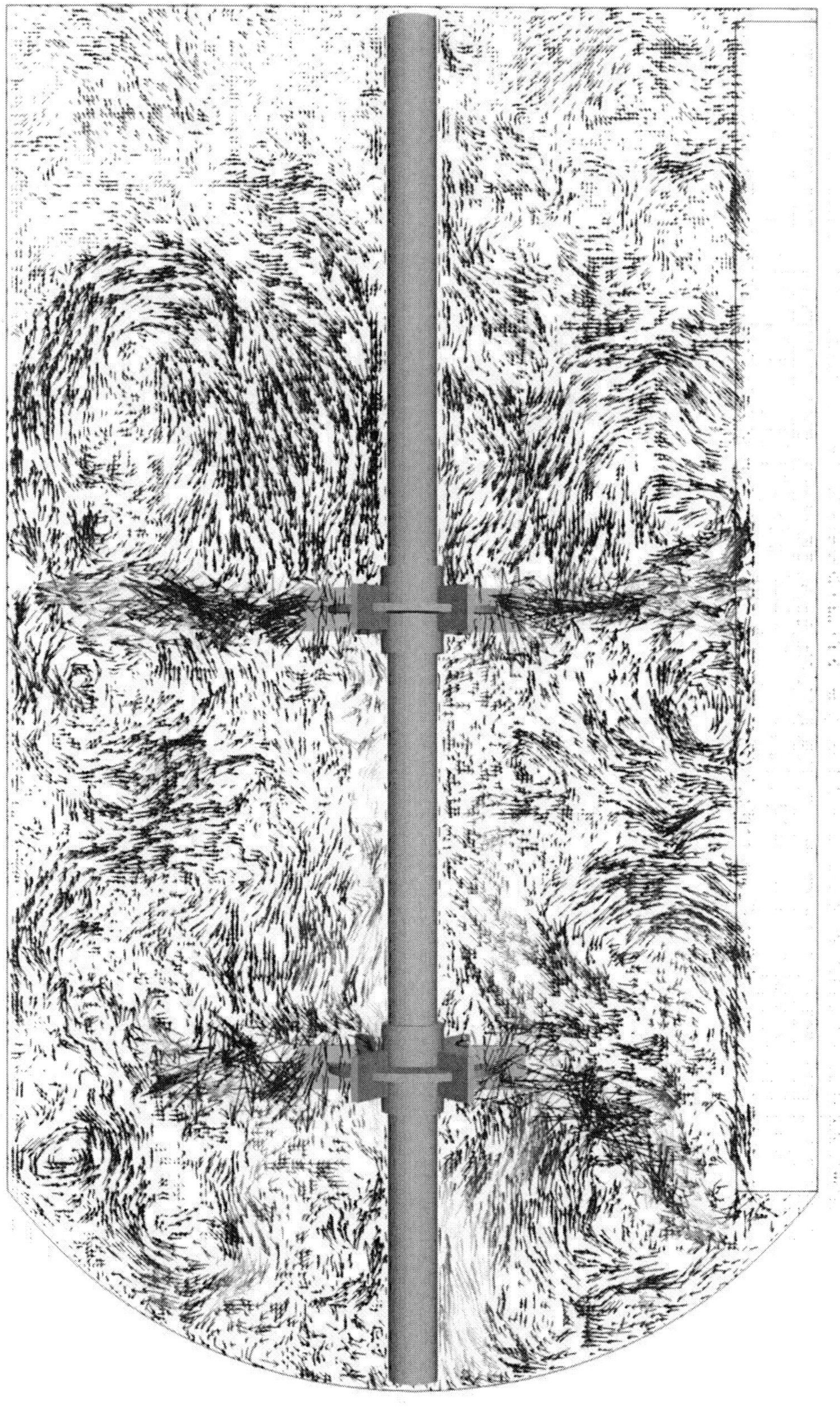

Figure 13: Velocity magnitude glyphs

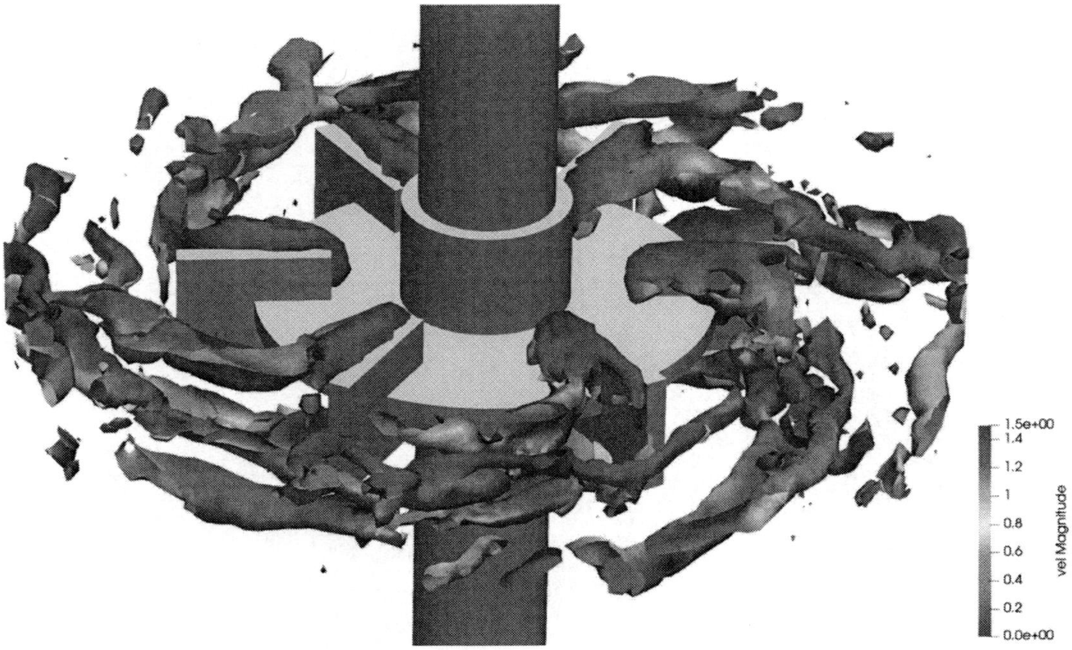

Figure 14: Characteristic divided rollup vortices

The compute details for the simulations are given in the following tables

Reactor with single Rushton impeller

Compute used (nodes)	2
Total memory used (GB)	9.67
Total real number of grid points (Million)	33.55
angular velocity (rpm)	2672
Resolution (mm)	0.625
time step (micro sec)	5.16
Total wall time (hours)	74
Number of rotations ran in 74 hours	200
Number of rotations ran in 1 hour	2.7

Reactor with double Rushton impellers

Compute used (nodes)	10
Total memory used (GB)	62.35
Total real number of grid points (Million)	210
angular velocity (rpm)	250
Resolution (mm)	0.36
time step (micro sec)	44.1
Total wall time (hours)	24
Number of rotations ran in 24 hours	34
Number of rotations ran in 1 hour	1.4

Conclusions

Lattice Boltzmann Method is presented as a promising alternative to solve complex turbulent fluid flows with moving geometries. New class of LB models with enhanced stability leading to turbulence model free simulations is presented in the form of Entropic Lattice Boltzmann Method. Flow in a single-impeller stirred tank reactor is simulated. Characteristic upper and lower circulation loops are captured. Radial, Tangential velocity profiles, TKE profiles are accurately predicted Flow in a double-impeller stirred tank reactor is simulated where velocity profiles in the vicinity of both the impellers accurately predicted. Velocity profiles along height of the tank at various radial positions predicted. Characteristic roll-up vortices are captured.

References

[1] F. Garcia-Ochoa, V.E. Santos, and E. Gomez. "Stirred Tank Bioreactors". In: *Comprehensive Biotechnology* (2011), pp. 179–198. DOI: 10.1016/b978-0-08-088504-9.00108-2.

[2] Maike Kuschel et al. "Validation of Novel Lattice Boltzmann Large Eddy Simulations (LB LES) for Equipment Characterization in Biopharma". In: *Processes* 9.6 (2021), p. 950. DOI: 10.3390/pr9060950.

[3] David Roush et al. "Toward in silico CMC: An industrial collaborative approach to model-based process development". In: *Biotechnology and Bioengineering* 117.12 (2020), pp. 3986–4000. DOI: 10.1002/bit.27520.

[4] Johannes Wutz et al. "Computational fluid dynamics (CFD) as a tool for industrial UF/DF tank optimization". In: *Biochemical Engineering Journal* 160 (2020), p. 107617. DOI: 10.1016/j.bej.2020.107617.

[5] Yogesh Waghmare et al. "Drawdown of floating solids in stirred tanks: Scale-up study using CFD modeling". In: *International Journal of Pharmaceutics* 418.2 (2011), pp. 243–253. DOI: 10.1016/j.ijpharm.2011.05.039.

[6] Cees Haringa, Ruben Vandewijer, and Robert F. Mudde. "Inter-compartment interaction in multi-impeller mixing: Part I. Experiments and multiple reference frame CFD". In: *Chemical Engineering Research and Design* 136 (2018), pp. 870–885. DOI: 10.1016/j.cherd.2018.06.005.

[7] Cees Haringa, Ruben Vandewijer, and Robert F. Mudde. "Inter-compartment interaction in multi-impeller mixing. Part II. Experiments, sliding mesh and large Eddy simulations". In: *Chemical Engineering Research and Design* 136 (2018), pp. 886–899. DOI: 10.1016/j.cherd.2018.06.007.

[8] Hugo Hartmann et al. "Assessment of Large eddy and RANS Stirred Tank Simulations by Means of LDA". In: *Chemical Engineering Science* 59 (June 2004), pp. 2419–2432. DOI: 10.1016/j.ces.2004.01.065.

[9] Sauro Succi. "The Lattice Boltzmann Equation - For Fluid Dynamics and Beyond". In: vol. 22. Feb. 2003, p. 368. DOI: 10.1016/S0997-7546(02)00005-5.

[10] Sauro Succi. "Lattice Boltzmann 2038". In: *EPL (Europhysics Letters)* 109 (Mar. 2015). DOI: 10.1209/0295-5075/109/50001.

[11] Iliya Karlin, A. Ferrante, and H. Öttinger. "Perfect entropy functions of the Lattice Boltzmann method". In: *EPL (Europhysics Letters)* 47 (Jan. 1999), p. 182. DOI: 10.1209/epl/i1999-00370-1.

[12] Sauro Succi, Iliya Karlin, and Hudong Chen. "Colloquium: Role of the H theorem in lattice Boltzmann hydrodynamic simulations". In: *Reviews of Modern Physics - REV MOD PHYS* 74 (Nov. 2002), pp. 1203–1220. DOI: 10.1103/RevModPhys.74.1203.

[13] Iliya Karlin et al. "Maximum Entropy Principle for Lattice Kinetic Equations". In: *Physical Review Letters* 81 (July 1998), pp. 6–9. DOI: 10.1103/PhysRevLett.81.6.

[14] Santosh Ansumali and Iliya Karlin. "Entropy Function Approach to the Lattice Boltzmann Method". In: *Journal of Statistical Physics* 107 (Apr. 2002), pp. 291–308. DOI: 10.1023/A:1014575024265.

[15] Santosh Ansumali, Iliya Karlin, and Hans Öttinger. "Minimal entropic kinetic models for hydrodynamics". In: *EPL (Europhysics Letters)* 63 (Sept. 2003), pp. 798–804. DOI: 10.1209/epl/i2003-00496-6.

[16] S. Chikatamarla, Santosh Ansumali, and Iliya Karlin. "Entropic Lattice Boltzmann Models for Hydrodynamics in Three Dimensions". In: *Physical review letters* 97 (Aug. 2006), p. 010201. DOI: 10.1103/PhysRevLett.97.010201.

[17] Praveen Kolluru et al. "Lattice Boltzmann model for weakly compressible flows". In: *Physical Review E* 101 (Jan. 2020). DOI: 10.1103/PhysRevE.101.013309.

[18] Manjusha Namburi, Siddharth Krithivasan, and Santosh Ansumali. "Crystallographic Lattice Boltzmann Method". In: *Scientific Reports* 6 (June 2016), p. 27172. DOI: 10.1038/srep27172.

[19] Siddharth Krithivasan, Siddhant Wahal, and Santosh Ansumali. "Diffused bounce-back condition and refill algorithm for the lattice Boltzmann method". In: *Physical Review E* 89.3 (2014). DOI: 10.1103/physreve.89.033313.

[20] Stanley Osher and James A Sethian. "Fronts propagating with curvature-dependent speed: Algorithms based on Hamilton-Jacobi formulations". In: *Journal of Computational Physics* 79.1 (1988), pp. 12–49. DOI: 10.1016/0021-9991(88)90002-2.

203b Predicting the Effect of Gradients on Cell Culture Performance in Large Scale Bioreactors

Katherine Raudenbush, John Thomas, Marianthi Ierapetritou

Scale-up of bioreactors is necessary for industrialization of monoclonal antibody production but can lead to the formation of spatial gradients in important culture parameters, including dissolved gases, metabolites, and pH. Different concentrations or values of these culture parameters can have adverse effects on cell culture dynamics, resulting in lower cell density, productivity, and product quality (N-linked glycosylation) [1,2]. Cells experience fluctuations of environmental conditions as they move throughout large scale bioreactors, exposing them to not only transient suboptimal conditions, but also oscillating conditions over time, which has known negative effects [3-6].

Modeling of integrated fluid dynamics and bio-phase kinetics in large scale bioreactors can be achieved using large eddy simulation (LES) computational fluid dynamic (CFD) models with Euler-Lagrange tracking of micro-organisms [7]. Incorporation of free-surface hydrodynamics, multiphase mixing, mass transport and reaction kinetics based on local concentrations enables in-silico modeling of process performance in large scale bioreactors. This framework is henceforth referred to as a CFD-kinetic model.

We present the application of the CFD-kinetic model on the production of monoclonal antibodies (mAbs) in Chinese hamster ovary (CHO) cells in large scale bioreactors operated in fed-batch mode. In fed-batch operation, spatial gradients are more prominent for dissolved gases supplied continuously in the form of sparged air, whereas gradients of metabolites are greatest temporally with spikes at each feed time over the 14-day process [8]. To reduce computation time and avoid unnecessary complexity when modeling the full fed-batch process, metabolite concentrations are considered homogeneous and constant in the CFD-kinetic model for short periods of time. Pseudo-steady spatial gradients of dissolved gases are resolved with the CFD-kinetic model at each day, with metabolite concentrations and cell densities at each day defined by kinetic models of CHO cell metabolism [9]. Predictions of viable cell density (VCD) and mAb production rates are compared to predictions based on spatially averaged tanks to predict the impact of gradients on process performance.

References

1. Brunner, M., et al., *Investigation of the interactions of critical scale-up parameters (pH, pO2 and pCO2) on CHO batch performance and critical quality attributes.* Bioprocess and Biosystems Engineering, 2017. **40**(2): p. 251-263.
2. Ivarsson, M., et al., *Evaluating the impact of cell culture process parameters on monoclonal antibody N-glycosylation.* Journal of Biotechnology, 2014. **188**: p. 88-96.
3. Anane, E., et al., *Scale-down cultivation in mammalian cell bioreactors—The effect of bioreactor mixing time on the response of CHO cells to dissolved oxygen gradients.* Biochemical Engineering Journal, 2021. **166**.
4. Brunner, M., et al., *The impact of pH inhomogeneities on CHO cell physiology and fed-batch process performance - two-compartment scale-down modelling and intracellular pH excursion.* Biotechnology Journal, 2017. **12**(7): p. 1600633.
5. Gao, Y., et al., *Combined metabolomics and proteomics reveals hypoxia as a cause of lower productivity on scale-up to a 5000-liter CHO bioprocess.* Biotechnology Journal, 2016. **11**(9): p. 1190-1200.
6. Serrato, J.A., et al., *Heterogeneous conditions in dissolved oxygen affect N-glycosylation but*

not productivity of a monoclonal antibody in hybridoma cultures. Biotechnology and Bioengineering, 2004. **88**(2): p. 176-188.
7. Haringa, C. *An analysis of organism lifelines in an industrial bioreactor using Lattice-Boltzmann CFD.* Engineering in Life Sciences, 2022. p.1-16.
8. Lara, A.R., et al., *Living With Heterogeneities in Bioreactors: Understanding the Effects of Environmental Gradients on Cells.* Molecular Biotechnology, 2006. **34**(3): p. 355-382.
9. Jimenez Del Val, I., et al., *Dynamics of immature mAb glycoform secretion during CHO cell culture: An integrated modelling framework.* Biotechnology Journal, 2016. **11**: p. 610–623.

203c Fluid-Dynamic Simulation of a Flat-Panel Bioreactor with Emphasis on Mixing, Mass Transfer and Inorganic CO_2 Chemistry

John Parra-Alvarez, Mauro Lua, Lieve Laurens, Hari Sitaraman, Jonathan Stickel, Heinnickel Mark, Paddock Troy

The potential for microalgae to have an impact on direct air capture increases as we better understand the driving mechanisms behind growth and fluid-dynamics in the different cultivation systems [1]. The design and operation of bioreactors affect the photosynthetic process of the microorganisms and effectively contribute to CO_2 capture [2]. Additionally, the biomass produced via growth can become an attractive raw material for the bio-feed/bio-fuel production [3,4]. Despite all these potential advantages, it is important to characterize the influence in growth and mass transfer of the availability of dissolved inorganic carbon, pH, reactor operation among others. This work is two-fold. First, we use computational fluid dynamics (CFD) to simulate the cultivation system, a flat panel bioreactor. We utilize the Euler-Euler multiphase framework from OpenFOAM to characterize the multiphase system including mass, momentum and energy between the gas and the liquid phases [5]. We analyze the impact of sparging system design on mixing and mass transfer by varying orifice distribution in the sparger manifold. Second, we implement the inorganic reaction rates for the CO_2 chemistry and analyze their impact on mass transfer at different growth pH. These trends would help us optimize the design and operation of the bioreactor for effective utilization of CO_2.

Bibliography.

- Cordoba-Perez, M.; De Lasa, H. CO2-derived carbon capture and photon absorption efficiency by microalgae in novel photo- BioCREC. *Eng. Chem. Res.* **2020**, *59*, 14710–14716.
- Adamczyk, M.; Lasek, J.; Skawinska, A. CO2 Biofixation and Growth Kinetics of Chlorella vulgaris and Nannochloropsis gaditana. *Biochem. Biotechnol.* **2016**, *179*, 1248–1261.
- Rawat, I.; Kumar, R.R.; Mutanda, T.; Bux, F. Biodiesel from microalgae: A critical evaluation from laboratory to large scale production. *Energy* **2013**, *103*, 444–467.
- Zhao, B.; Su, Y. Process effect of microalgal-carbon dioxide fixation and biomass production: A review. *Sustain. Energy Rev.* **2014**, *31*, 121–132.
- Rahimi, M., Sitaraman, H., Humbird, D. and Stickel, J., Computational fluid dynamics study of full-scale aerobic bioreactors: Evaluation of gas–liquid mass transfer, oxygen uptake, and dynamic oxygen distribution. *Chemical Engineering Research and Design*, **2018,** 139: 283–295.

203d Comparison of Modeling Methodologies for Simulating a Gassed-Agitation System with Experimental Validation

Christopher Tyler, John Thomas

Aerobic and microaerobic fermentation processes require significant aeration for proper organism performance in both stirred tanks and air-lift columns. In stirred tanks, the power requirement for agitation is an important design parameter that indirectly relates back to bulk mass transfer coefficient kLa. Recent work has shown that simulations predict kLa and that breakup and coalescence are essential for predicting the agitator power in gassed stirred tanks. Three methods are available for modeling gas/liquid flow in an agitated vessel: full-resolution of the gas/liquid interface; lagrangian particle tracking of all bubbles; and a lagrangian-parcel approach, where groups or parcels of bubbles are modeled collectively. For the later two methods, there is in addition a choice of how to model breakup and coalescence. This work will compare simulation and experimental results to assess how well these various methods predict the power in a gassed-agitated system, with a specific emphasis on the impeller flooding.

203e Geometrical Optimization of Side-Entry Propeller Designs for the Homogenization Process in Biogas Fermenters

Markus Kolano, Matthias Kraume

Mixing in biogas fermenters is of high importance for the stability of the process: Local gradients in temperature, concentration as well as stratification effects, i.e. swimming and sinking layers, have to be minimized. All this has to be achieved at very low power inputs to sustain an economic process, which is especially difficult considering the distinct non-Newtonian rheology of most biogenic substrates.

The most common mixing technology in biogas industry is the mechanical agitation with one or multiple side-entry propellers. To deal with the above-described complex agitation problem, a large number of different propeller designs has been developed in industry. This complicates the choice of an efficient agitator for a plant operator since very limited data are available for a specific propeller geometry.

The abundance of propeller geometries results from the application of varying design principles. Oftentimes propellers are developed using rules for marine propulsion, e.g. These designs lead to an inefficient mixing process, though, as is shown in this presentation by decolorization experiments conducted in a pilot-scale fermenter ($V_R = 724$ L, s. Fig. 1).

Based on these mixing experiments, fundamental rules are developed in this talk for an efficient propeller design. These rules are then applied in CFD studies, in which various geometrical propeller parameters are tested for their effect on efficiency. Using 3D printed models, these numerically derived relationships are then also validated experimentally in mixing and PIV experiments in pilot-scale.

It can be shown that the efficiency of a propeller for side-entry mixing applications is described by the constant $c_{Eff} = F_{Ax}\, \rho^{-1}\, n^{-1}\, d^{-1}\, P^{-0.5}$, where F_{Ax} is axial thrust, ρ density, n stirring rate, and P power input. The number of propeller blades is of minor importance for the efficiency of a propeller geometry (s. Fig. 2 left; higher c_{Eff} denote more efficient designs). A pitch angle of $\alpha \approx 40°$ is most efficient; this angle should be applied at the blade tips to achieve a high pumping rate to power input ratio (s. Fig, 2 right; a blade twist of 20° fulfills this criterion). Compared to flat blades, the use of a suitable hydrofoil design can increase efficiency by more than 20%.

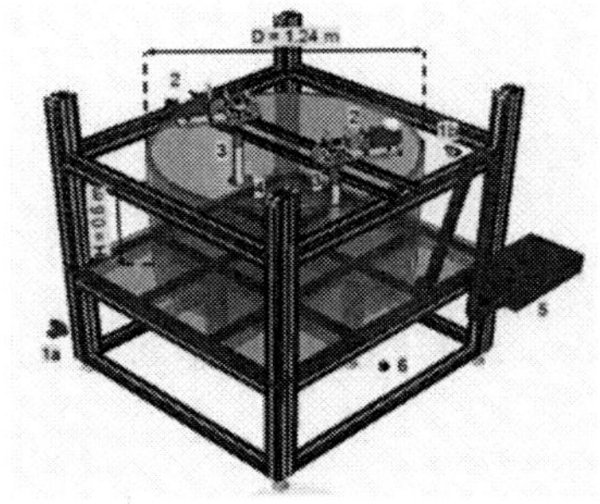

Figure 1: Pilot-scale fermenter; 1a+b: cameras to monitor decolorization experiments, 2: motors, 3: strain gauges to measure axial thrusts, 4: propellers, 5: PIV laser, 6: cameras for PIV measurements

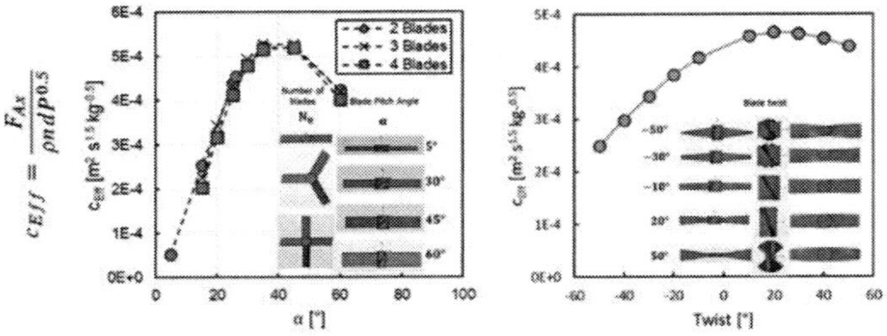

Figure 2: Numerically derived efficiency of some propeller designs: number of blades and pitch angle in the left, various blade twists on the right

203f Multiphysics Simulation of an Internal Loop Airlift Photobioreactor for Microalgae Cultivation

Lifeng Li, Zulfida Mohamad Hafis Mohd Shafie, Ting-Yi Huang, Raymond Lau, Chi-Hwa Wang

Microalgae can be used for the production of high quality single cell protein. The cultivation of microalgae can be accomplished in a photobioreactor where the requisite culturing conditions such as light intensity, CO_2, O_2, pH, and temperature, are satisfied [1,2]. In order to optimize the reactor geometrical configurations for achieving the maximal biomass yield, a multiphysics model is developed in this study which integrates the hydrodynamics, radiative transfer and algal growth kinetics using computational fluid dynamics (CFD) simulations.

Transient, two-dimensional, Eulerian Euler-Euler approach with the standard k-$\mathcal{E}$ turbulence model is developed for studying the flow dynamics of the gas-liquid or gas-liquid-solid flows in a concentric-tube internal loop airlift reactor (see Fig. 1) using ANSYS Fluent. The radiative transfer model approximates the light intensity as a function of the medium depth and the dry biomass concentration, based on experimentally measured data. The microalgae cell growth kinetic model is developed based on experimentally measured data which considers the effect of light intensity on the algal growth rate. Based on the developed multiphysics model, we evaluate the growth of microalgae as affected by the reactor geometrical parameters, particularly the reactor scale. This work paves the way for developing a low-cost, high-production, pilot-scale modular photobioreactor system suitable for commercial deployment.

References:

[1] J. Pruvost, F. L. Borgne, A. Artu, J. Cornet, J. Legrand, Industrial photobioreactors and scale-up concepts, in: Advances in Chemical Engineering, Vol. 48, 2016, pp. 257–310.

[2] L.-L. Zhuang, D. Yu, J. Zhang, F.-F. Liu, Y.-H. Wu, T.-Y. Zhang, G.-H. Dao, H.-Y. Hu, The characteristics and influencing factors of the attached microalgae cultivation: a review, Renewable and Sustainable Energy Reviews 94 (2018) 1110–1119.

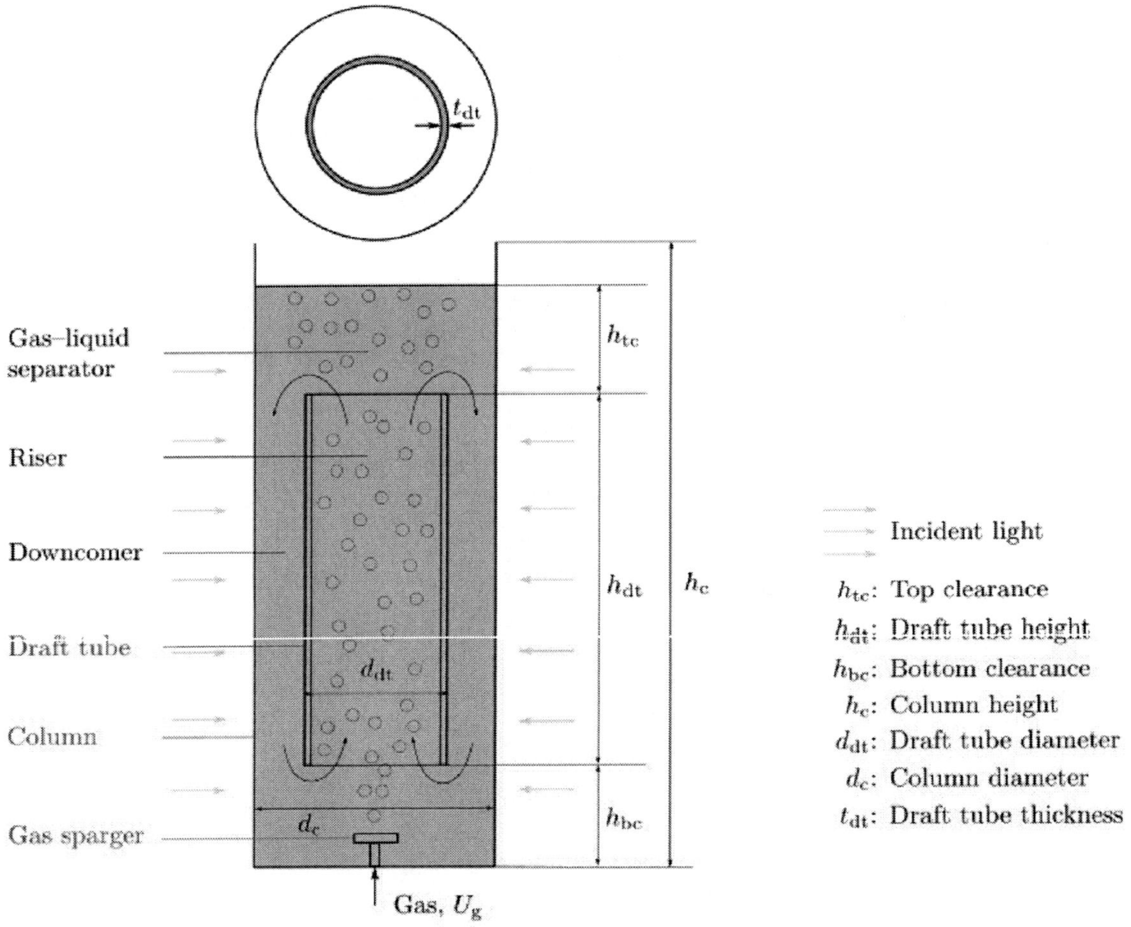

Gas–liquid separator
Riser
Downcomer
Draft tube
Column
Gas sparger
t_{dt}
h_{tc}
h_{dt}
h_{bc}
h_c
d_{dt}
d_c
Gas, U_g
Incident light
h_{tc}: Top clearance
h_{dt}: Draft tube height
h_{bc}: Bottom clearance
h_c: Column height
d_{dt}: Draft tube diameter
d_c: Column diameter
t_{dt}: Draft tube thickness

267a On the Dispersion Dynamics of Liquid-Liquid Surfactant-Laden Flows in Static Mixers

Juan Valdes, Fuyue Liang, Lyes Kahouadji, Seungwon Shin, Jalel Chergui, Damir Juric, Omar K. Matar

The characterization of liquid-liquid mixing processes is arguably one of the most chal- lenging problems of modern process engineering, encompassing a broad range of non- idealities such as complex interfacial dynamics, surfactant-laden flows, among several others. These operations are heavily involved in a wide spectrum of applications, ranging from fast-moving consumer goods in the food, personal care and pharmaceutical indus- tries to high-end sectors such as (petro) chemical, polymer processing, and biotechnology [Paul et al., 2004]. In particular, the dispersion dynamics behind static mixers has shown growing interest in the past decades, given their potential as an attractive alternative to conventional stirred vessels [Valdes et al., 2022]. Despite this, design and scale-up procedures for these processes are largely dependant on limited empirical models and heuristics due to the intrinsic complexity of liquid-liquid systems. In turn, this has led to over-design and over-consumption practices aiming to ensure a minimum threshold of product quality, thus compensating for the lack of fundamental knowledge on the mixing itself [Valdes et al., 2022]. Blind application of simplified models and design heuristics on inherently complex flow systems can pose numerous risks to the manufacturing process, resulting in significant disruptions such as major adjustments to the plant, inflated costs due to over-sizing, and lost opportunity of unsuccessful products due to lacking quality or inadequate physico-chemical properties [Valdes et al., 2022, Paul et al., 2004].

The effects of surfactants are a critical area of interest when dealing with liquid-liquid systems since the properties and stability of most structured goods (i.e., emulsion-based products) are heavily correlated with the manipulation of the interfacial tension between phases [Wong et al., 2015, Leal-Calderon et al., 2007, Valdes et al., 2022]. Very few studies dealing with static mixers have considered the influence of adding surfactants on the mix- ing performance, measured from the droplet size distribution (DSD) generated, and none of them have done so from a fundamental stance. Experimental studies [Lobry et al.,2011, Das et al., 2013, Barega et al., 2013, Farzi et al., 2016] have demonstrated that higher surfactant concentrations (hence lower interfacial tension) results in smaller DSDs. This is often attributed to higher breakage and lower coalescence events, which stems from decreased internal restoring forces and increased repulsion between drops [Barega et al., 2013, Valdes et al., 2022]. However, a deep dive into the physical mechanisms affecting fundamental processes such as droplet deformation, breakage and coalescence under the influence of surface-active agents has not been considered thus far. Further- more, a thorough evaluation of different relevant surfactant properties (e.g., solubility, elasticity, etc) on the dispersion dynamics has not been conducted either. Although some ground has already been covered in regard to the general trends expected, continuation on similar empirical lines of research will not be sufficient to further advance this field [Hakansson, 2019, Valdes et al., 2022]. Considering this, the present work seeks to provide a deeper understanding of the underlying governing physical mechanisms dictating immiscible liquid-liquid mixing when handling surfactant-laden flows.

In this study we implement high-fidelity, three-dimensional direct numerical simulations coupled with a state-of-the-art code based on a hybrid front-tracking level-set interface capturing algorithm, mounted on a massively parallelized computer architecture [Shin et al., 2017, 2018]. This hybrid formulation solves explicitly the unsteady dynamics of the free-interface, providing a wealth of information and detail on interfacial dynamics and physical phenomena which are inaccessible experimentally or via volume-averaged numerical approaches. Based on our previous and on-going studies, we consider a SMX-type mixer with two different initial configurations for the dispersed phase: 1) isolated cases: individual droplets (1 and 3 drops), mimicking a controlled syringe injection; and 2) pre-mixed cases: multiple droplets with different sizes simulating a pre-dispersed inlet. With these configurations we explore two main surfactant flows in order to isolate the effects of different properties on the static mixer's performance: insoluble and soluble surfactants. For the insoluble case, we vary the elasticity parameter β, which relates with the strength of the surfactant in terms of its influence on the interfacial tension. In the soluble cases, different levels of desorption (measured from the Bi number) and adsorption (ka) capabilities are explored and contrasted against the clean and fully insoluble case.

The surfactant-free simulations elucidate a two-step dispersion process taking place in the static mixer, consisting of an initial elongational deformation with no breakup, mostly driven by uniform extensional stresses, followed by a myriad of interfacial instabilities which results in breakup events via Rayleigh-Plateau or end-pinching mechanisms. These mechanics can be closely related with different metrics such as the maximum stretching efficiency [Liu et al., 2005] or the flow topology parameter [Soligo et al., 2020]. A similar two-step dispersion process is observed for the surfactant-laden cases. Nonetheless, a significant growth of the interfacial area and number of daughter droplets is noted for the insoluble cases. Furthermore, a saturation point was detected at increasing surfactant strength in the insoluble case, but with different outcomes when comparing a single droplet (lower interfacial growth at higher β) vs. a low coalescing three drop set-up (higher growth at higher β). Simulations involving soluble surfactants are still on-going but it is expected to observe the formation of surfactant-covered immobile regions and nearly surfactant free mobile regions [Batchvarov et al., 2020]. The implications of these gradients on the restoring and disruptive stresses acting on the droplets will be explored.

References

Esayas W. Barega, Edwin Zondervan, and Andre B. De Haan. Influence of physical properties and process conditions on entrainment behavior in a static-mixer settler setup. Industrial and Engineering Chemistry Research, 2013. ISSN 08885885. doi: 10.1021/ie301580m.

A. Batchvarov, L. Kahouadji, M. Magnini, C. R. Constante-Amores, S. Shin, J. Chergui, D. Juric, R. V. Craster, and O. K. Matar. Effect of surfactant on elongated bubbles in capillary tubes at high Reynolds number. Physical Review Fluids, 5(9), 9 2020. ISSN 2469-990X. doi: 10.1103/PhysRevFluids.5.093605.

Mainak D. Das, Andrew N. Hrymak, and Malcolm H.I. Baird. Laminar liquid-liquid dis- persion in the SMX static mixer. Chemical Engineering Science, 2013. ISSN 00092509. doi: 10.1016/j.ces.2013.06.047.

G. A. Farzi, N. Rezazadeh, and A. Parsian Nejad. Droplet Formation Study in Emulsification Process by KSM using a Novel In Situ Visualization System. Jour- nal of Dispersion Science and Technology, 37:575–581, 2016. ISSN 15322351. doi: 10.1080/01932691.2015.1052144.

Andreas Hakansson. Emulsion Formation by Homogenization: Current Understanding and Future Perspectives. Annual Review of Food Science and Technology, 10:239–258, 2019. ISSN 19411421. doi: 10.1146/annurev-food-032818-121501.

Fernando Leal-Calderon, Jerome Bibette, and Veronique Schmitt. Emulsion science: Basic principles. Springer, 2nd edition, 2007. ISBN 0387396829. doi: 10.1007/ 978-0-387-39683-5.

Shiping Liu, Andrew N. Hrymak, and Philip E. Wood. Drop breakup in an SMX static mixer in laminar flow. Canadian Journal of Chemical Engineering, 2005. ISSN 00084034. doi: 10.1002/cjce.5450830501.

Emeline Lobry, Felicie Theron, Christophe Gourdon, Nathalie Le Sauze, Catherine Xuereb, and Thierry Lasuye. Turbulent liquid-liquid dispersion in SMV static mixer at high dispersed phase concentration. Chemical Engineering Science, 2011. ISSN 00092509. doi: 10.1016/j.ces.2011.06.073.

Edward L. Paul, Victor A. Atiemo-Obeng, and Suzanne M. Kresta. Introduction. In Handbook of Industrial Mixing - Science and Practice, chapter 1, pages xxxiii–lx. John Wiley andamp; Sons, 2004. ISBN 0-471-26919-0.

Seungwon Shin, Jalel Chergui, and Damir Juric. A solver for massively parallel direct numerical simulation of three-dimensional multiphase flows. Journal of Mechanical Sci- ence and Technology, 31(4), 4 2017. ISSN 1738-494X. doi: 10.1007/s12206-017-0322-y.

Seungwon Shin, Jalel Chergui, Damir Juric, Lyes Kahouadji, Omar K. Matar, and Richard V. Craster. A hybrid interface tracking – level set technique for multiphase flow with soluble surfactant. Journal of Computational Physics, 359:409–435, 4 2018. ISSN 00219991. doi: 10.1016/j.jcp.2018.01.010.

Giovanni Soligo, Alessio Roccon, and Alfredo Soldati. Effect of surfactant-laden droplets on turbulent flow topology. Physical Review Fluids, 5(7):073606, 7 2020. ISSN 2469- 990X. doi: 10.1103/PhysRevFluids.5.073606.

Juan P. Valdes, Lyes Kahouadji, and Omar K. Matar. Current advances in liquid–liquid mixing in static mixers: A review. Chemical Engineering Research and Design, 177, 1 2022. ISSN 02638762. doi: 10.1016/j.cherd.2021.11.016.

S. F. Wong, J. S. Lim, and S. S. Dol. Crude oil emulsion: A review on formation, classification and stability of water-in-oil emulsions. Journal of Petroleum Science and Engineering, 135:498–504, 2015. ISSN 09204105. doi: 10.1016/j.petrol.2015.10.006.

267b Numerical Simulation of Surfactant-Laden Emulsion Formation in a Stirred Vessel

Fuyue Liang, Juan Valdes, Lyes Kahouadji, Seungwon Shin, Jalel Chergui, Damir Juric, Omar K. Matar

Given the heavy involvement of immiscible liquid mixing both in our daily life (e.g., food and cosmetics) and in high-end industrial applications (e.g., drug development), understanding the interfacial flows under highly dynamic condition (e.g., within a stirred vessel) is of practical significance. The addition of surface-active agents (surfactants), either as additives or as contaminants, can affect both the mixing performance and the final quality of the mixed product. Under dynamic flow conditions, surfactants not only reduce the interfacial tension but also modify the interfacial rheology (although this will not be considered here) in a complex manner that depends on the surfactant mass transfer between bulk-phase and interface, and its transport along the interface. The elevated complexity hampers the understanding and control of interfacial behaviours within a surfactant-laden mixing vessel, which is a prevalent scenario in industry.

Previous studies have shown the deviation of predicted drop size in surfactant-laden mixing vessel by using surfactant-free drop breakup model [1]. Subsequent studies were dedicated to investigating the surfactant effect on the final drop size and their distribution [2] by considering surfactant concentration [3, 4], rheological properties of surfactant-laden interfaces [5], and types of surfactant head group [6]. More related work has been on the modification of the velocity field in stirred vessels due to the presence of surfactant [7, 8]. However, these studies only addressed the problem regarding the drop size distribution inside a surfactant-laden mixer leaving a gap in our understanding of the underlying physical mechanisms governing the interfacial behaviours within such a system. Nevertheless, interfacial dynamics on the surfactant-laden interface has been well studied in the early times by investigating drop deformation subjected to a simple flow field [9], resembling linear shear flow and uniaxial elongational flow. More recently, numerical studies have been carried out on a surfactant-laden drop dispersed in turbulent channel flow [10]. From the short review above, a systematic study of flow structure and interfacial dynamics in a practical mixing unit, i.e., stirred vessel, in the presence of surfactant remains lacking.

The present study aims to establish the transparent interplay among interfacial deformation, surfactant transportation, and the underlying flow structure inside a cylindrical stirred vessel equipped with a pitched blade turbine (PBT). Massively parallel, three-dimensional, interface-tracking, large eddy simulations [11] of O/W emulsification are deployed to provide detailed, realistic visualization of the intricate interfacial dynamics coupled to the turbulent flow fields, from the onset of impeller rotation through to the attainment of a dynamic steady state. In particular, this study investigates the effect of surfactant elasticity (the sensitivity of interfacial tension to the surfactant concentration), the Biot number (the ratio of characteristic desorptive to convective time scale), and Marangoni stresses. Allied to this, the transient drop counts, and their size distribution have been tracked.

The simulations have elucidated that the impeller rotation introduces a primary vortex (generating a centripetal force at the interface) which leads to a high surfactant concentration region near the axis and a low concentration area near the vessel wall. The continuous phase appears to be characterized mainly by pure rotational flow region (flow topology parameter [12], $Q = -1$) and pure elongational flow region ($Q = +1$), with small regions of pure shear flow ($Q = 0$) squeezing in between. Ligaments are produced as the deforming interface reaches the impeller, followed by elongational deformation and pinch-off giving rise to small drops. The ligament stretching is shown to be mostly located in pure shear flow regions, while drop breakup follows the interaction with the rotational flow (turbulent vortices). Moreover, the simulations have demonstrated that the presence of surfactant is associated with interfacial rigidification, suppression of end-pinching [13], and promotion of tip-streaming. The interfacial dynamics in the surfactant-laden system alters the evolution of drops count and their size distribution, where the appearance of the first dispersed drop occurs earlier and a larger number of relatively small drops are observed in comparison to their surfactant-free counterparts carried out in our previous work.

REFERENCES

[1] **Koshy, A., Das, T. R., and Kumar, R.** "Effect of surfactants on drop breakage in turbulent liquid dispersions." *Chemical engineering science*, **43**(3), 649-654 (1988).

[2] **Groeneweg, F., Van Dieren, F., and Agterof, W. G. M.** "Droplet break-up in a stirred water-in-oil emulsion in the presence of emulsifiers." *Colloids and Surfaces A: Physicochemical and Engineering Aspects*, **91**, 207-214 (1994).

[3] **El-Hamouz, A.** "Effect of surfactant concentration and operating temperature on the drop size distribution of silicon oil water dispersion." *Journal of Dispersion Science and Technology,* **28**(5), 797-804 (2007).

[4] **Chatzi, E. G., Boutris, C. J., and Kiparissides, C.** "On-line monitoring of drop size distributions in agitated vessels: 2. Effect of stabilizer concentration". *Industrial andamp; engineering chemistry research,* **30**(6), 1307-1313 (1991).

[5] **Lucassen-Reynders, E. H., and Kuijpers, K. A.** "The role of interfacial properties in emulsification." *Colloids and Surfaces*, **65**(2-3), 175-184 (1992).

[6] **Goloub, T., and Pugh, R. J.** "The role of the surfactant head group in the emulsification process: Single surfactant systems." *Journal of colloid and interface science*, **257**(2), 337-343 (2003).

[7] **Arora, K., Sureshkumar, R., Scheiner, M. P., and Piper, J. L.** "Surfactant-induced effects on turbulent swirling flows." *Rheologica acta*, **41**(1-2), 25-34 (2002).

[8] **Goloub, T., and Pugh, R. J.** "The role of the surfactant head group in the emulsification process: Single surfactant systems." *Journal of colloid and interface science*, **257**(2), 337-343 (2003).

[9] **Eggleton, C. D., Tsai, T. M., and Stebe, K. J.** "Tip streaming from a drop in the presence of surfactants." *Physical review letters*, **87**(4), 048302 (2001).

[10] **Soligo, G., Roccon, A., and Soldati, A.** "Breakage, coalescence and size distribution of surfactant-laden droplets in turbulent flow." *Journal of Fluid Mechanics*, **881**, 244-282 (2019).

[11] **Shin, S., Chergui, J., Juric, D., Kahouadji, L., Matar, O. K., and Craster, R. V.** "A hybrid interface tracking–level set technique for multiphase flow with soluble surfactant." *Journal of Computational Physics*, **359**, 409-435 (2018).

[12] **Soligo, G., Roccon, A., and Soldati, A**. "Effect of surfactant-laden droplets on turbulent flow topology." *Physical Review Fluids*, **5**(7), 073606 (2020).

[13] **Constante-Amores, C. R., Kahouadji, L., Batchvarov, A., Shin, S., Chergui, J., Juric, D., and Matar, O. K.** "Dynamics of retracting surfactant-laden ligaments at intermediate Ohnesorge number." *Physical Review Fluids*, **5**(8), 084007 (2020).

267c The Transition to Aeration in Turbulent Two-Phase Mixing in Stirred Vessels

Lyes Kahouadji, Fuyue Liang, Juan Valdes, Seungwon Shin, Jalel Chergui, Damir Juric, Omar K. Matar

Flow mixing inside stirred vessels occurs in a large array of industrial applications and produces complex dynamical structures. These structures, such as those seen in [1] for single-phase flow, exert a strong influence on the mixing efficiency. Many fast-moving consumer goods involve the manufacturing of so-called structured products (e.g., foods, creams, detergents), which are mass-produced via complex multiphase mixing processes of several base products, commonly carried out in agitated tanks. Some viscous products require rapid mixing, but the appearance of bubbles can lead to undesirable partial bottle-filling and process inefficiencies. In contrast, for other processes, such as those used in ice cream manufacturing (involving non-Newtonian fluids and emulsifiers [2,3] and those that deploy bioreactors, the promotion of 'aeration' is essential. Thus, it is crucial to predict the mixing patterns in stirred vessels, and to demarcate the aeration threshold as a function of relevant system parameters, such as fluid properties, and impeller geometry and rotational speed.

Mixing that induces aeration in a stirred vessel is perhaps one of the most challenging configurations to predict via accurate computational methods. This difficulty is due firstly to the high-density gap between the two phases (air-liquid mixing), which makes solvers difficult to converge. Moreover, the presence of a myriad interfacial singularities, such us interfacial breakup and coalescence, produces high velocity gradient, reduces consequently the computational time-step. Therefore, the time consuming for the computation is very long. Furthermore, an accurate calculation, with almost any assumption, it requires many computational skills that is not easy to combine together such as multiphase flows, fluid-structure interaction, turbulence modelling.... Finally, the multiscale dynamical (shear-layers and vortical structures) also requires important resolution. Consequently, it becomes very challenging and very costly in term of computational method to predict accurate mixing inducing aeration, despite the enormous industrial applications.

The combination of innovative numerical algorithms in a single numerical framework, able to accurately handle coupled physics in fluid mechanics such as multiphase interface motion, fluid-structure interaction and turbulence modelling, is essential for predicting complex mixing flow in stirred vessels. Moreover, a robust high-performance computing architecture enables in-depth understanding of previously inaccessible physics for such extreme flow regimes. In the context of aeration due to mixing in a stirred vessel, where the density ratio between the air and the liquid is $O(10^3)$, it is crucial to provide a robust and reliable numerical framework able to encompass all the techniques listed above.

We consider the mixing of a viscous fluid by the rotation of a pitched blade turbine inside an open, cylindrical tank, with air as the lighter fluid above. To examine the flow and interfacial dynamics, we utilise a highly-parallelised implementation of a hybrid front-tracking/level-set method that employs a domain-decomposition parallelisation strategy. Our numerical technique is designed to capture faithfully complex interfacial deformation, and changes of topology, including interface rupture and dispersed phase coalescence. As shown via transient, three-dimensional direct numerical simulations, the impeller induces the formation of primary vortices that arise in many idealised rotating flows as well as several secondary vortical structures resembling Kelvin-Helmholtz, vortex breakdown, blade tip vortices, and end-wall corner vortices. As the rotation rate increases, a transition to aeration is observed when the interface reaches the rotating blades leading to the entrainment of air bubbles into the viscous fluid and the creation of a bubbly, rotating, free surface flow. The mechanisms underlying the aeration transition are probed as are the routes leading to it, which are shown to exhibit a strong dependence on flow history.

Before reaching the challenging bubbly aeration regime, there are many attempts of computing mixing flow with many assumptions, For example, Ciofalo et al.[4] performed a three-dimensional turbulent flow simulation, where the flow equations are in the rotating reference frame of the impeller with the addition of a conventional linear logarithmic "wall function' as in [5]). The work presented in [6] compared alternative computational methods: the

first replaced the impeller by suitable boundary conditions, and the second consisted of dividing the computational domain into two concentric and partially-overlapping parts; the inner region, containing the impeller, where the flow is simulated in the impeller rotating reference frame, while in the outer region, simulations are conducted in the laboratory reference frame. This technique requires information exchange between the two regions. More recently, the work presented in [7] used the lattice Boltzmann method to simulate the flow in aerated bioreactors, and [8] coupled a volume-of-fluid method to a Reynolds stress model to capture the gas-liquid interface and turbulent flow agitated by pitched blade turbines where the interface deflection reached the impeller hub.

To the best of our knowledge, studies involving unsteady, turbulent, and high deformable free surface flows have been restricted to situations wherein the interface deflection does not descend beyond the impeller blades. As a result, these studies are unable to analyse, in detail, the rich and complex vortical structures accompanying such flows. Furthermore, the phenomenon of aeration has not yet been studied in detail via numerical simulations despite its obvious importance to industrial applications as highlighted above. Aeration involves the development of sufficiently large interfacial deformations that lead to the interaction of the free surface with the rotating impeller. This, in turn, brings about the entrainment and dispersion of the lighter phase into the underlying denser phase; for gas (air)-liquid systems, the dispersed phase corresponds to bubbles (of air).

Our aim in this work is to present the intricacies of two-phase mixing flows in a stirred vessel via a standard LES (Large Eddy Simulation) Smagorinsky-Lilly turbulence model coupled with a Direct-Forcing Method for the motion of the impeller [9,10]. Moreover, our numerical framework is formulated in the context of a high-fidelity front-tracking technique for the interface [11,12] which is able to handle complex interfacial deformation, pinchoff and coalescence. Applying this new numerical scheme for stirred vessels, our results elucidate, for the first time, the challenging transition to aeration and its dependence on flow history (e.g., ramping up impulsively from a stationary state vs. increasing the impeller rotational speed following the achievement of a steady-state at lower speeds). A crucial metric for mixing-induced aeration that a numerical framework should be capable of furnishing is the gas/air holdup which corresponds to the relative amount of gas/air held in the liquid phase. We provide results for the temporal variation of the holdup from our simulation data.

REFERENCES

[1] Batels, C., Breuer, M., Wechdler, K., andamp; Durst, F. (2020). Computational fluid dynamics applications on parallel-vector computer: computations of stirred vessel flows. Computer andamp; Fluids 31, 69–97.

[2] McClements, D. J. (2004). Food Emulsions: Principles, Practices, and Techniques 2nd Edition.

CRC Press.

[3] Douglas Goff, H. (1997). Colloidal aspects of ice cream - A review. Int. Dairy J. 7, 363–373.

[4] Ciofalo, M., A. Brucato, A., Grisafi, F., andamp; Torraca, N. (1996). Turbulent flow in closed and free-surface unbaffled tanks stirred by radial impellers. Chemical Engineering Science 51(14), 3557–3573.

[5] Launder, B. E. andamp; Spalding, D. B.(1974). The numerical computation of turbulent flows. Comp. Meth. Appl. Mech. Engineering 3, 260–289.

[6] Brucato, A., Ciofalo, M., Grisafi, F., andamp; Micale, G. (1998). Numerical prediction of flow fields in baffled stirred vessels: A comparison of alternative modelling approaches. *Chemical Engineering Science 53*(21), 3653-3684.

[7] Witz, C., Treffer, D., Hardiman, T., andamp; Khinast, J. (2016). Local gas holdup simulation and validation of industrial-scale aerated bioreactors. *Chem. Eng. Sci. 152*, 636–648.

[8] Li, L., Wang, J., Feng, L. andamp; Gu, X. (2017). Computational fluid dynamics simulation of hydrodynamics in an uncovered unbaffled tank agitated by pitched blade turbines. *Korean J. Chem. Eng. 34*(11), 2811–2822.

[9] Fadlun, E. A., Verzicco, R., Orlandi, P., andamp; Mohd-Yusof, J. (2000). Combined immersed-boundary finite-difference methods for three-dimensional complex flow simulations. *J. Comp. Phys. 161*, 35–60.

[10] Mohd-Yusof, J. (1997). Combined immersed-boundary / b-spline methods for simulations of flow in complex geometries. *Center for Turbulence Research Annual Research Briefs* 317–327.

[11] Shin, S., Chergui, J. andamp; Juric, D. (2017). A solver for massively parallel direct numerical simulation of three-dimensional multiphase flows. *J. Mech. Sci. Tech. 31*, 1739.

[12] Shin, S., Chergui, J., Juric, D., Kahouadji, L., Matar, O.K. andamp; Craster, R. V. (2018). A hybrid interface tracking-level set technique for multiphase flow with soluble surfactant. *J. Comp. Phys. 359*, 409–435.

267d Multiscale Modelling of Food Emulsions: From Molecules to Mixing Equipment

Marco Ferrari, Alessio Lombardo Pontillo, Antonio Buffo, Gianluca Boccardo, Marco Vanni, Daniele Marchisio

Mayonnaise is a stable liquid-liquid, oil-in-water, emulsion with high content of the disperse phase. The production of mayonnaise is a typical mixing process, in which the ingredients (i.e. egg yolk, vinegar, oil and water) are first mixed together in large stirred vessels at moderate rotational speed. Then the formed emulsion is finally fluxed into a high-shear mixer, where the oil droplets undergo breakage until the final size distribution is reached. This last step is crucial to fine tune the droplet size distribution, in order to determine the properties of the final product, such as structure, stability over time, taste and colour. In this work, we aim to model the last step of the production process by means of a Computational Fluid Dynamics approach coupled with a Population Balance Model (CFD-PBM) and with a Dissipative Particle Dynamics (DPD) model. A schematic representation of the investigated apparatus, so called cone mill mixer (1), is reported in Figure 1. It is constituted of a solid conical frustum rotor inside a slightly larger stator of the same shape, forming a small gap in which the emulsion flows (as it can be seen in Figure 1) and experiences high-shear rates, due to the high rotational speed of the rotor. To properly describe the non-Newtonian dynamics of the emulsion, the fluid is considered as a shear thinning pseudo single phase. In order to describe the evolution of the droplet size distribution, the Population Balance Equation is employed (2), in which coalescence and breakage of oil droplets are taken into account by appropriate coalescence and breakage kernels, which depend on local flow conditions. These kernels are also function of the local concentration of the surface active molecules contained in mayonnaise, namely phospholipids and apoproteins. In order to understand the role of these molecules in stabilizing the oil-water interface and with the final objective of improving the coalescence and breakage kernels currently employed in CFD-PBM simulations, a molecular modelling technique is used here for the first time in this context. Molecular Dynamics (MD) is employed to build a coarse-grained engineering model, based on DPD, to describe the molecular interactions between phospholipids and apoproteins and the oil-water interface and extract useful information, that can be in turn used in the development of coalescence and breakage kernels. MD simulations are run with the software GROMACS, whereas DPD simulations are run with the commercial software CULGI-Siemens. CFD-PBM simulations are performed with the open-source code OpenFOAM (version 6.0).

The flow field show particular patterns in agreement with experimental data and other numerical simulations (3). CFD-PBM results are reported in terms of the mean oil droplets diameter in Figure 2, along a longitudinal section of the cone mill. As expected, the mean droplets size decreases along the flow direction, since droplets undergo breakage induced by the high-shear rates inside the mixer.

This effort is carried out in the context of the VIMMP project (www.vimmp.eu), where the entire workflow will serve to devise a marketplace for generic multiscale and multiphysics simulations. The VIMMP project has received funding from the European Union's Horizon 2020 Research Innovation Programme under Grant Agreement n. 760907.

References

(1) A. Dubbelboer, J. J. M. Janssen, H. Hoogland, E. Zondervan, J. Meuldijk, Chem. Eng. Sci. 148 (2016) 32–43.

(2) D. Li, Z. Gao, A. Buffo, W. Podgorska, D. L. Marchisio, AIChE J. 63 (2017) 2293–2311.

(3) X. Li, J. Zhang, L. Xu, J. Hydrodyn. 26 (2014) 431–435.

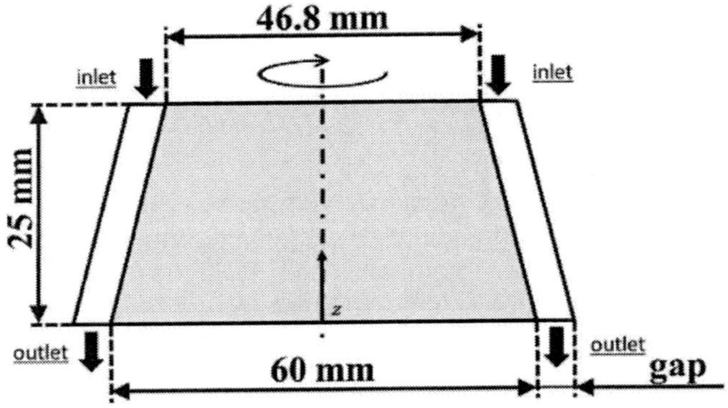

Figure 1 Sketch of the cone mill mixer.

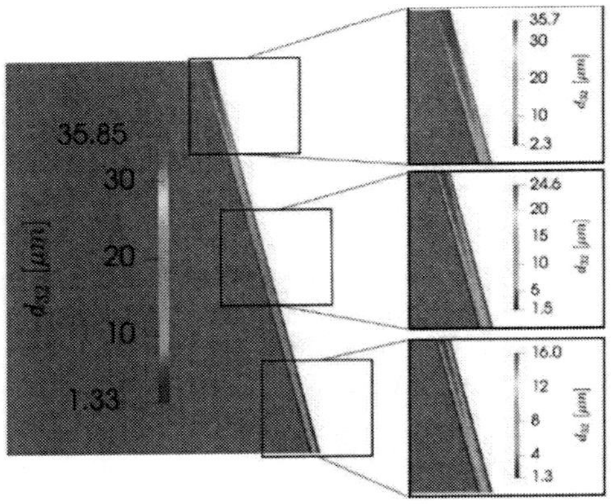

Figure 2 Trend of the mean oil droplets diameter along a longitudinal section of the cone mill mixer.

267e Rapid Mixing in Microfluidic Devices with Induced-Charge Electro-Osmosis for Improved Control over Precipitation Reactions

Dishika Gupta, Baggie W. Nyande, Kiran Mathew Thomas, Fei Li, Andrew T.C. Mak, Richard Lakerveld

Precipitation involves the rapid formation of particles, which is frequently used in various chemical industries. Product quality attributes such as the particle size distribution often need to be controlled tightly to optimize downstream processes and to attain a desired final product functionality. However, control of product quality is often difficult to achieve in precipitation processes due to the entangled kinetics of mixing, nucleation, growth and aggregation at usually high supersaturation levels[1], [2]. The particles may nucleate and grow before complete mixing is achieved, which can lead to a non-uniform product quality that is difficult to predict. Thus, achieving rapid mixing in the order of the induction time for nucleation of the particles is needed to create uniform and predictable conditions for precipitation[3]. Micromixers are attractive for precipitation processes as they offer better reproducibility and controlled reaction environments compared to conventional stirred tank reactors[4]. Microfluidic devices usually need to be integrated with either passive or active mixing techniques to avoid relying on slow diffusive mixing only in the laminar flow regime. Passive mixers exploit favorable flow conditions created by specific structural designs, whereas active mixers employ external energy fields such as magnetic, electric, or pressure from, for example, ultrasound to manipulate mixing during operation[5].

Electrokinetically driven active micromixers utilizing induced-charge electro-osmosis (ICEO) have gained attention due to their ability to provide rapid and uniform mixing at relatively low AC voltages[6]. The working principle is based on the application of an electric field that drives double layer formation and a non-linear fluid flow near a polarizable conducting surface by creating microvortices[7]. Apart from mixing applications, ICEO has been utilized for fluid pumping, particle sorting and separation process in microfluidic devices[8]–[10]. Moreover, other electric-field-assisted mixing methods such as AC-electrothermal[11] and electrohydrodynamic flows[12] have been explored for mixing of multiple fluid streams in nanoparticle synthesis and liposome synthesis, respectively. Among these electric-field-assisted mixing techniques, ICEO is an attractive choice, as it can provide efficient mixing without requiring any electrical conductivity gradients or high voltage amplitudes[13], [14]. However, use of ICEO for improved control over precipitation processes has not been reported yet.

The objective of this work is to design and characterize a microfluidic device to obtain improved control over precipitation processes through rapid mixing based on ICEO-driven flows. A triple-electrode design was used to enhance mixing through the formation of asymmetric microvortices. The design was analogous to a field effect transistor with a floating gate electrode in the center[15]. CFD simulations were used to scale up the device so that clogging during the continuous precipitation processes was avoided while maintaining good mixing performance. The CFD simulations also revealed that the mixing length required to achieve complete mixing was drastically reduced in the presence of ICEO. Next, the mixing performance of the device was experimentally characterized for different fluid flow rates and electrical field parameters such as fluid conductivity, field frequency, voltage amplitude, and the phase difference between the driving electrodes for the precipitation of silver chloride that was adopted as a model system. The ICEO-based micromixer provided best mixing for fluids with low electrical conductivity when antiphase voltage was applied to the driving electrodes at an intermediate AC-frequency. Furthermore, the supersaturation profiles, mixing lengths and mixing patterns inside the microfluidic device were determined numerically at the optimal mixing conditions, which were compared with the mixing conditions without ICEO. The particle size distribution of the silver chloride nanoparticles was characterized at different mixing conditions, flow rates, and supersaturation ratios. The silver chloride nanoparticles produced with ICEO-driven mixing were smaller and exhibited a narrower size distribution compared to the corresponding case without ICEO under all the tested conditions. The ability to make small particles of uniform size demonstrates the good potential of the mixer for precipitation processes. Finally, the feasibility of the concept for application to a cocrystallization system was demonstrated for the carbamazepine-succinic acid cocrystal model system. Cocrystallization systems are an attractive class of applications due to their low conductivities. The yield of the cocrystals was higher when using ICEO, which was attributed to the improved mixing.

Acknowledgement: The work described in this abstract was supported by a grant from the Research Grants Council of the Hong Kong Special Administrative Region, People's Republic of China (Project No. 16214418).

References

[1] O. Söhnel and J. Garside, "Precipitation: Basic Principles and Industrial Applications," *Butterworth-Heinemann Oxford,* 1992.

[2] A. Mersmann and M. Kind, "Chemical engineering aspects of precipitation from solution," *Chem. Eng. Technol.*, vol. 11, no. 1, pp. 264–276, 1988.

[3] H. C. Schwarzer and W. Peukert, "Experimental investigation into the influence of mixing on nanoparticle precipitation," *Chem. Eng. Technol.*, vol. 25, no. 6, pp. 657–661, 2002.

[4] V. Hessel, S. Hardt, and H. Löwe, *Chemical Micro Process Engineering Microreactors Microsystem Technology Catalytic Membranes and Membrane Reactors*. 2004.

[5] V. Hessel, H. Löwe, and F. Schönfeld, "Micromixers - A review on passive and active mixing principles," in *Chemical Engineering Science*, 2005, vol. 60, no. 8-9 SPEC. ISS., pp. 2479–2501.

[6] M. Z. Bazant, "Nonlinear Electrokinetic Phenomena," in *Encyclopedia of Microfluidics and Nanofluidics*, D. Li, Ed. Boston, MA: Springer US, 2008, pp. 1461–1470.

[7] T. M. Squires and M. Z. Bazant, "Induced-charge electro-osmosis," *J. Fluid Mech.*, vol. 509, no. 509, pp. 217–252, 2004.

[8] J. S. Paustian, A. J. Pascall, N. M. Wilson, and T. M. Squires, "Induced charge electroosmosis micropumps using arrays of Janus micropillars," *Lab Chip*, vol. 14, no. 17, pp. 3300–3312, 2014.

[9] Y. Ren *et al.*, "Scaled particle focusing in a microfluidic device with asymmetric electrodes utilizing induced-charge electroosmosis," *Lab Chip*, vol. 16, no. 15, pp. 2803–2812, 2016.

[10] K. Zhang, Y. Ren, L. Hou, X. Feng, X. Chen, and H. Jiang, "An efficient micromixer actuated by induced-charge electroosmosis using asymmetrical floating electrodes," *Microfluid. Nanofluidics*, vol. 22, no. 11, pp. 1–11, 2018.

[11] H. Sun, Y. Ren, Y. Tao, T. Jiang, and H. Jiang, "Three-Fluid Sequential Micromixing-Assisted Nanoparticle Synthesis Utilizing Alternating Current Electrothermal Flow," *Ind. Eng. Chem. Res.*, vol. 59, no. 27, pp. 12514–12524, 2020.

[12] P. Modarres and M. Tabrizian, "Electrohydrodynamic-Driven Micromixing for the Synthesis of Highly Monodisperse Nanoscale Liposomes," *ACS Appl. Nano Mater.*, vol. 3, no. 5, pp. 4000–4013, 2020.

[13] Z. Wu and D. Li, "Micromixing using induced-charge electrokinetic flow," *Electrochim. Acta*, vol. 53, no. 19, pp. 5827–5835, 2008.

[14] C. C. Chang and R. J. Yang, "Electrokinetic mixing in microfluidic systems," *Microfluid. Nanofluidics*, vol. 3, no. 5, pp. 501–525, 2007.

[15] Y. Wu, Y. Ren, Y. Tao, L. Hou, Q. Hu, and H. Jiang, "A novel micromixer based on the alternating current-flow field effect transistor," *Lab Chip*, vol. 17, no. 1, pp. 186–197, 2017.

564a North American Mixing Forum Award Lecture: The Progression of Computational Fluid Dynamics As a Tool for Process and Equipment Design in Mixing

Eric E. Janz

In industrial equipment design, tools used are often a balance of cost and accuracy. This is true of both experimental and computational tools. Computational Fluid Dynamics (CFD) has been a tool used by the mixing industry since the early 1990s in various forms. Early practical applications in rotating mixing used extensive symmetry assumptions and experimental impeller discharge boundary conditions to allow for practicable insights to be gained. These tools were an excellent example of George E. P. Box's adage, "All models are wrong, but some are useful." The simulations were not "right," but they were useful, allowed for quick optimization, and were an excellent supplement to experimental work.

As computers became more powerful and simulation codes became more sophisticated, so did the ability to use CFD as a tool. By the mid-2000s, 3D simulations using CAD geometries were commonplace for more complicated applications. For large, complex projects, they offered insights that were difficult, expensive, or nearly impossible to extract experimentally.

While large eddy simulations (LES) and Lattice Boltzmann method (LBM) techniques were used for many years in academic and nation-state scale applications, by 2015, they had become tractable for industrial-scale applications. This allowed time-accurate physics to be simulated at scales and flows that could mimic industrial-scale applications. Phenomena such as time-accurate transient forces on mechanical components (fluid-structure interactions), reactions, and surface processes could be simulated in a time frame that would be immediately useful.

Looking ahead, a continued combination of lower cost and faster computational capabilities, along with fast-growing technologies like machine learning, will accelerate the use of CFD as a primary design and troubleshooting tool. Ultimately this will allow for industrially tractable, fully turbulent, transient simulations, including wall-controlled phenomena (e.g., heat transfer), to be completed and even coupled with mechanical analyses. As these tools develop further, more and more of the applications of CFD will be performed by practitioners rather than modeling experts.

This talk will look back at several snapshots in time to review the impact of the simulation tools of the time, reflect on the current state of the art and look forward to some of the next significant impacts of CFD. It will also examine mixing tasks where CFD can be used as a tool to easily, accurately, and effectively capture needed engineering answers and examine tasks that currently require users to be modeling and mixing experts.

406a Development of Turbulence Measurement Technique Using Electrical Resistance Tomography for Multiphase Flows

Weiguo Xie

Turbulence affects every aspect of our life, from atmospheric and oceanic flows, to transport of fluids all the way to a multitude of essential industrial processes. However, it has always been an enormous scientific challenge to quantify turbulence for multiphase flows due to their complex features and the lack of diagnostic tools, especially when the system is of three phases (solid, liquid and gas) such as the pulp phase in flotation cells employed for minerals recovery. Turbulence plays an important role in the flotation process since it affects the three main sub-processes: air dispersion, particle suspension and bubble-particle collection interactions. Entrainment of fine particles is also affected by turbulence. Quantification methods for turbulent flows require the measurement of the fluctuating (mass, constituent's phase, heat, momentum, energy) transport properties at a sufficiently high spatial and temporal resolution. Electrical resistance tomography (ERT) has been applied to a wide range of scientific and engineering fields such as geophysical study, clinical diagnosis, chemical or mineral process measurement and optimization. When air bubbles in flotation cells move through the ERT sensor plane, the conductivity distribution fluctuates with turbulence. By measuring these fluctuations, turbulence information can be extracted by using a Green-Kubo method. ERT has been applied to measure turbulence profiles in a 60-liter air/water flotation cell, a $3m^3$ air/water flotation cell and a Metso $3m^3$ industrial flotation cell (as shown in Figure 1). The results showed agreement with turbulence measurement technique using piezoelectric vibration sensor. This ERT turbulence measurement technique for the multiphase flows can be potentially used to wider industrial applications.

Metso 3 m^3 RCS flotation test rig

ERT sensor

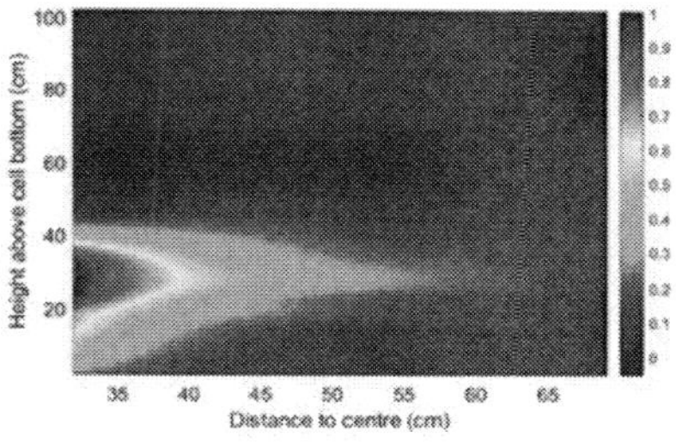

Turbulence profile measured by ERT

Figure 1: Turbulence measurements for Metso 3 m^3 RCS flotation test rig using ERT

406b Power, Flow and Efficiency: Effect of Adding Tiplets on Impeller Efficiency

Richard Grenville, Jason Giacomelli, Benjamin Boyer, Bernd Gigas

Impellers in stirred tanks can be considered as pumps and their performance can be assessed in terms of their power drawn, flow generated, and efficiency. The hydraulic efficiency is defined as the kinetic energy of the flowing fluid divided by the mechanical energy input by the impeller, or the mass flow efficiency is the mass of fluid pumped per unit of energy input by the impeller. Both definitions show that hydrofoils are the most efficient impellers and Rushton turbines, the least (Grenville et al., Chem. Engg., 2017).

Impellers also generate a trailing vortex at their tip where the high-pressure region on the front of the blade meets the low pressure on the back. This is the location of the highest energy dissipation rate within the vessel where break-up processes, such as and the dispersion of immiscible liquids, are controlled. But the energy dissipated in the trailing vortex is not available to generate flow, and there is evidence that suppressing the vortices with tiplets at the end of the impeller blades does increase their efficiencies. This result has been found for the Lightnin A6000 (with tips) and A6100 and the Philadelphia Mixing Solutions MHE and MHET (with tips) hydrofoils.

It is not clear that the increased efficiency results in, for example, an increase in the rate of blending, reducing the blend time, or a reduction in the power required to suspend particles in a slurry. Blending and solids suspension performance for impellers with and without tiplets will be compared and analyzed in terms of their power input and flow generation.

406c Hydrodynamics Investigation of the USP Dissolution Testing Apparatus 1 (Basket Apparatus) Filled with 500 Ml Media Volume Using Particle Image Velocimetry (PIV)

Chadakarn Sirasitthichoke, Piero Armenante

Drug dissolution testing is an analytical tool used to assess the *in vitro* dissolution performance of pharmaceutical oral solid dosage forms. The USP Apparatus 1 (basket apparatus) is a common tool used for this purpose. Any changes in the system can impact the system hydrodynamics and cause variability of results, thus affecting product quality. The literature on the hydrodynamics is Apparatus 1 is quite limited. However, recently our group quantified the velocity distribution using standard basket 40-mesh size in a vessel filled with 900-mL (Sirasitthichoke et al., Intern. J. Pharmaceutics:X 3 (2021) 100078). No such studies exist on the hydrodynamics of dissolution vessels filled with 500 mL, which is a common alternative to the 900 mL case. Here, we experimentally generated detailed maps of the flow field in the standard USP Apparatus 1 for a liquid fill volume of 500 mL using Particle Image Velocimetry (PIV) for different basket rotational speeds. The flow field was dominated by the tangential velocity component while the magnitudes of the axial and radial velocity components varied with location but were always low. A nearly horizontal small jet was seen emanating radially near the top edge of the basket. Velocities typically scaled well with increasing rotational speed in most regions of the vessel except for a region directly below the basket. A comparison of these results with those for the 900 mL fill volume case showed that the flows through the baskets were similar in both systems. The velocity distributions were also found to be similar, except in the region above the basket, which was affected by the presence of a small radial jet with an orientation significantly different between the 500-mL and the 900-mL systems. We expect that these results will be of significant relevance to dissolution testing scientists and practitioners. Even more importantly, assuming that the dissolution process occurs under sink conditions in both cases and that the tablet does not immediately disintegrate but instead erodes slowly, our results imply that *the amount of drug released in the dissolution media increases linearly with time. This in turn implies that the drug dissolved over time is independent of the liquid medium volume at least during the initial phase of the dissolution process* when the tablet surface area does not change significantly.

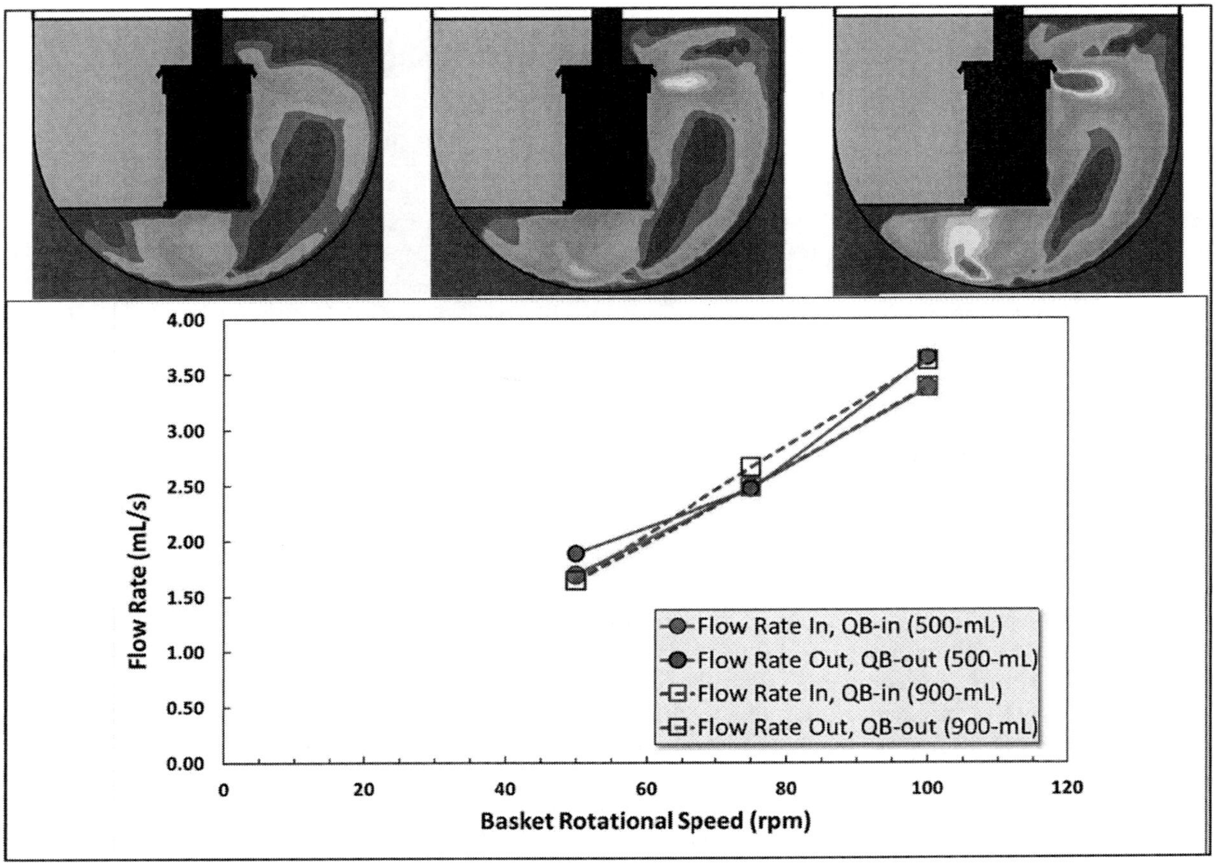

4.00
3.50
3.00
2.50
2.00
1.50
1.00
0.50
0.00
Flow Rate (mL/s)
0
20
40
60
80
100
120
Basket Rotational Speed (rpm)
Flow Rate In, QB-in (500-mL)
Flow Rate Out, QB-out (500-mL)
Flow Rate In, QB-in (900-mL)
Flow Rate Out, QB-out (900-mL)

406d Application of a Combined PIV/PTV-Method to Analyze Suspensions in (non-)Newtonian Media in Stirred Tanks

Markus Kolano, Matthias Kraume

Suspending solid particles in stirred tanks is a standard process and is therefore widely used. While for Newtonian liquids this process has already been extensively investigated, there is no comparable knowledge and database for solid suspension in non-Newtonian liquids.

Non-Newtonian flow properties are expressed, for example, in a viscosity that is dependent on shear rate. In polymer systems, e.g., additional elastic flow properties lead to the development of normal stress differences. Both properties influence the interaction between particles and fluid and thus also the suspension process in the stirred tank.

In order to quantify these effects, refractive index-matched solid-liquid systems are analyzed in this talk in a standard, flat bottomed acrylic glass tank (D=H=0.29) equiped with 6-bladed pitched blade turbine (s. Fig. 1). Glass spheres with a uniform diameter are used as solids. As liquid, an (aqueous) inverted sugar solution is used, which is additionally adjusted to the refractive index of the solids via salts. Non-Newtonian flow properties are created by the addition of polymers.

This system enables the simultaneous recording of velocities of the liquid phase via PIV and solid phase via Particle Tracking Velocimetry (PTV) in the laser cutting plane (s. Fig. 1). For PIV, small density-adapted tracer spheres, which reflect the laser light, are used. They can be selectively recorded with a band-pass filter with a wavelength range adjusted to the laser light (s. Fig. 2). PTV of the solid spheres is enabled by addition of rhodamine-6G: this colorant leads to fluorescence of the liquid in the laser cutting plane, which is selectively recorded using a long-pass filter (s. Fig. 2). The shadows caused by the solids in this plane then can be evaluated by PTV algorithms (s. Fig. 3).

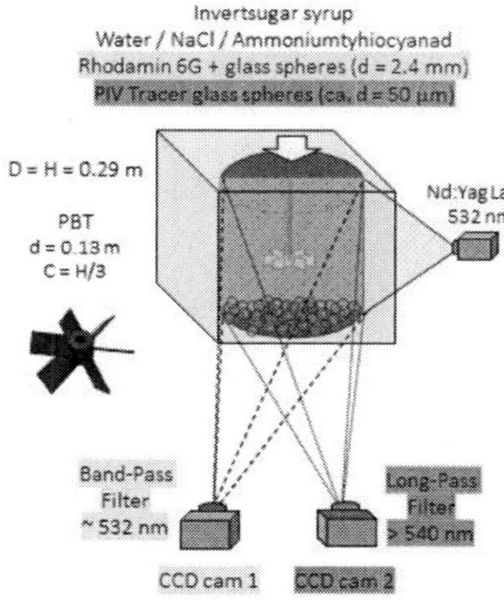

Figure 1: Experimental setup. For PIV, small tracer glass spheres and a band-pass filter are used; for PTV, a fluorescent colorant in combination with a long-pass filter and larger particles is used

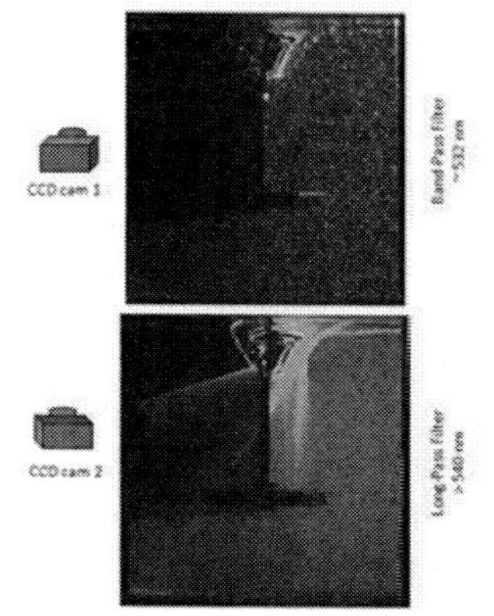

Figure 2: Sample images simultaneously acquired for both cameras / filters

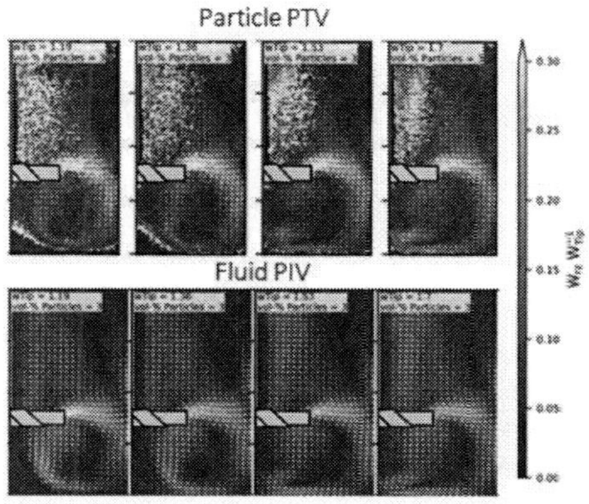

Figure 3: Sample velocity fields obtained using PIV and PTV algorithms

406f Experimental Investigation on the Power Input of Viscoelastic Fluids in an Unbaffled Tank

Alexander Maywurm, Lutz Böhm, Matthias Kraume

Stirring processes involving viscoelastic fluids are relevant in food industry, polymer industry and biotechnology, e.g., in xanthan fermentation, polystyrene production or fermentation of substrate in biogas processes. The DFG project ''Mixing of fluids with viscoelastic flow behaviour in stirred tanks'' aims to find basic principles for the efficient design of stirring processes involving viscoelastic fluids. The present research examines the power input of viscoelastic fluids in an unbaffled tank with various agitator geometries.

Material and methods

Power input experiments are conducted in two geometrically similar unbaffled flat-bottom cylindrical tanks (diameter T=480mm, T=290mm and T=90mm) equipped with either a three-stage cross-beam agitator (vertical of inclined blades), a blade or a grid agitator. The torque is recorded with a 20Nm and an 1Nm strain gauge-based torque sensor in the two larger scales. An Anton Paar RheolabQC fitted with stereolithographic miniature agitators is used for the scale-down experiments. Stereo particle image velocimetry (PIV) in the intermediate scale is selectively applied to help with the interpretation of the data obtained from power input measurements.

Shear-thinning viscoelastic aqueous polymer solutions of xanthan gum, carboxymethyl cellulose, polyacrylamide and polyethylene oxide at different concentrations are used to cover a wide range of rheological properties. Additionally, constant viscosity elastic fluids, so-called Boger fluids, based on glucose syrup are investigated to experimentally separate shear thinning effects from the influence of elasticity. All fluids are rheologically characterized in terms of flow curve, first normal stress differences, storage and loss module in an Anton Paar MCR302 rotary rheometer with a cone-plate geometry.

Results

Based on the results obtained with Boger fluids compared to the Newtonian reference, elasticity itself is found to have practically no influence on the power input in the laminar regime. Shear-thinning fluids with pronounced elastic however may show a significant increase in power input compared to weakly elastic fluids that is indirectly related to elasticity. The PIV measurements show that elastic forces weaken the secondary flow and thereby reduce the local shear rates near the agitator (figure 1). This results in an increase of the effective viscosity. The Metzner-Otto constant may reduce as much as two thirds when comparing to the other extreme case.

A sharp elasticity-induced increase of the power input is manifested in the transitional regime where Boger fluids deviate from the Newtonian reference and approach a constant power number that is concluded to form an elastic limit as indicated in figure 2. A similar increase in power input is found for shear-thinning elastic fluids with a steep dependence of the first normal stress difference on the shear rate. When inertia forces dominate the fluid flow a decrease in power input below the Newtonian reference can be observed which is contributed to the shear-thinning characteristic.

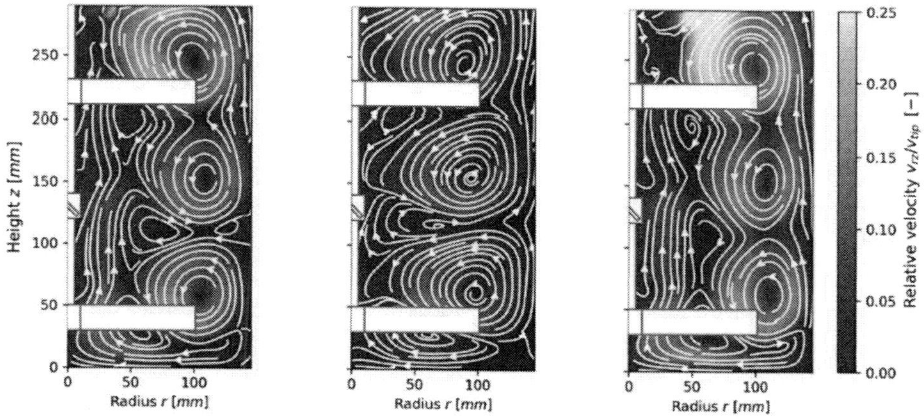

Figure 1:Secondary flow field of a three-stage cross-beam agitator with inclined blades.
A CMC solution at Re=300 exhibits reduced radial and vertical velocities and compartimization compared to the Newtonian reference (left at Re=200 and right at Re=360).

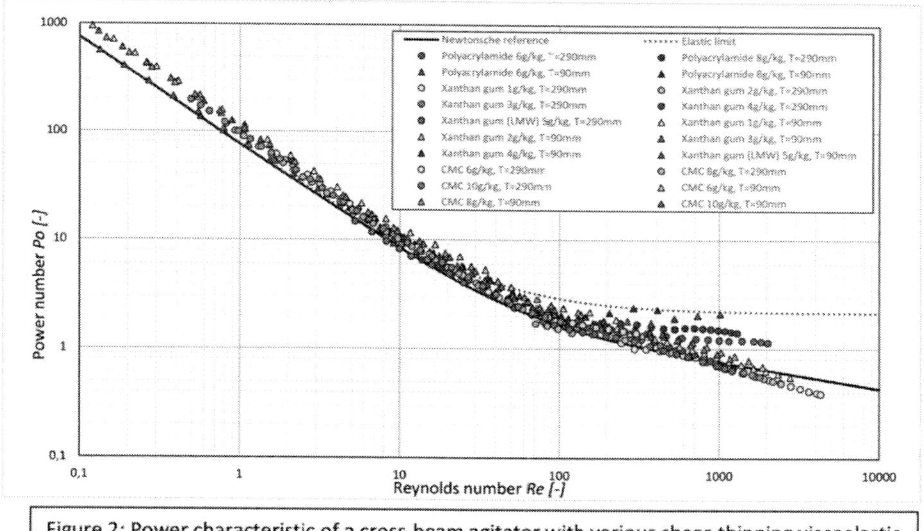

Figure 2: Power characteristic of a cross-beam agitator with various shear-thinning viscoelastic fluids calculated using a Metzner-Otto constant of $k_{MO} = 17$. The power number increases with elasticity in the laminar regime. In the transitional regime the data is scattered between the Newtonian reference and an Elastic limit found for Boger fluids.

511g Influence of mixing on nanoparticle size distribution

Faiz Khan, Xiangming Gu, Nicholas Brunelli, Barbara E. Wyslouzil, Jessica Winter

Polymer nanoparticles have been widely used as carriers for encapsulated hydrophobic drug molecules. Such drug delivery nanocarriers provide therapeutic efficacy by overcoming biological barriers and improving circulation time in the body. Nanoparticle size and particle size distribution affected by process synthesis conditions play a crucial role in determining therapeutic efficacy. Polymer nanoparticles synthesized through common nanoprecipitation techniques undergo rapid self-assembly in milliseconds. However, the mixing process in large batch reactors is accomplished in the order of minutes. This difference between the polymer self-assembly time and reactor mixing time results in a broad distribution of nanoparticles, with a high polydispersity. A polydisperse nanoparticle distribution compromises the uniformity of size-dependent properties, undesired for drug delivery applications. Hence, the goal is to achieve a reactor mixing time equivalent to the polymer self-assembly timescale. It is difficult to achieve rapid mixing in large-scale reactor vessels. Thus, microreactor designs such as T-mixer, Y-mixer, and confined impinging jet mixer have been used for polymer nanoparticles synthesis. Such dual inlet microreactors designs fail to achieve effective mixing under asymmetric flow conditions, i.e., when the flow rates of two inlets are unequal. This significantly restricts exploration of reaction parameters, such as supersaturation achieved in nanoprecipitation and limits applicability for processes involving expensive reagents, such as drugs. Our strategy involves decreasing the reactor length scale to allow for a smaller reactor volume and thus, achieve rapid mixing while being capable of operating under asymmetric flow conditions. Here, we present a three-inlet microreactor (also known as the Jet mixing reactor, JMR) that comprises two oppositely placed fluid streams that orthogonally impinge on a single fluid stream, developing vortexes and achieving rapid mixing in milliseconds. The proposed JMR resembles a cross-flow geometry and has the size of a dollar coin with inlet diameters ranging from 0.5-1 mm. As polymer nanoprecipitation kinetics is strongly affected by mixing inside the reactor, it is imperative to characterize mixing in the JMR. Mixing in the JMR was characterized using a competitive chemical reaction set, known as the Villermaux-Dushman reaction. Mixing dependence on different factors was explored, including dependence on inlet fluid stream flow rates, fluid viscosity, and cross-flow fluid ratios. Flow dynamics were also visualized using computational fluid dynamics with simulations on the COMSOL software. The competitiveness of JMR over conventional large batch vessels was assessed through model nanoparticle (Polylactic acid-co-glycolic acid, PLGA) synthesis, with nanoparticle size and particle size distribution serving as an evaluation criterion.

511c Solid-Liquid Flow and Mixing in Fluidic Oscillator

Ketan Madane, Vivek V. Ranade

Introduction

Fluidic oscillators (FOs) originated in the early 60s as a replacement for the then-infant stage transistors and logic circuits[1]. Since its inception, FOs have found a variety of like flow meters, actuators, sensors, liquid-liquid extractors, emulsions, and wastewater treatment. FOs have reported having excellent mixing characteristics owing to which they have been reported to be used for process and fine chemical engineering applications[2, 3]. Recently Pandit and Ranade[4] and Yang Yu et al.[5] have demonstrated the use of FO in continuous antisolvent crystallization of Paracetamol. Experimental and numerical characterizations of flow in FO have mainly been based on single-phase in past studies. The crystallization process generates solid particles resulting in solid-liquid flow in the FO. Therefore evaluating solid-liquid flow in FO becomes essential for establishing FO as a good crystallizer. Madane et al.[3] have computationally investigated an improved version of FO for single-phase flow, mixing, residence time distribution, and heat transfer characteristics. However, extending this computational modelling for solid-liquid flow and quantifying solid-liquid flow, solids residence time distribution and mixing in the presence of solids will be certainly valuable in further developing and optimizing FO geometry for crystallization. Solid-liquid flow in FOs has not been reported extensively in the open literature. In this work, we have both experimentally and computationally studied and quantified the solid-liquid flow dynamics in the FO. The current study focuses on (i) Experimental validation of the modelled solid-liquid flow and mixing in the FO through high-speed photography, (ii) Quantification of solid residence time and mixing performance in the presence of solids, and (iii) Studying the effect of the particles size on solid residence time and flow dynamics.

Experiments and Modelling

The solid-liquid flow in the FO was modelled with the help of the Eulerian-Lagrangian approach using Discrete particle modelling (DPM) in Ansys Fluent 2021 R2. In this numerical study, Paracetamol (ρ = 1263 kg/m^3) was selected as the discrete solid phase to mimic an actual model crystallization system. Single-sized spherical particles of diameters (D_p) of 50, 100, 150, and 200 µm were modelled. Two-way coupling of the continuous and discrete phases was employed. The turbulence in the FO was modelled by using the SST k−ω model with water as a continuous phase (ρ = 1000 kg/m^3, μ = 0.001 kg/m-s). High quality and a well-structured grid was used for the numerical solution to ensure the maximum possible accuracy of the solution (Figure 1). The simulations were performed in an unsteady mode owing to the unsteady periodic pattern of the oscillating jet. The particles were traced every time step that was set at s. This work will investigate the effects of particle size on axial dispersion, residence times, and velocities of particles. The effect of particles on the jet oscillation frequency will also be determined to quantify the effect of particles on the overall performance of the FO.

To investigate the solid-liquid flow in the FO, coloured particles of known sizes (distribution) will be injected along with the continuous flow of water via a T-junction. FO used for this experiment will be made of transparent material to allow visibility of the injected particles. The particles will be traced with a high-speed camera. The obtained video will be post-processed using the image analysis technique. By image analysis, the stagnation zones for the particles will be identified to determine solid-dead zones. The concentration of the particles will also be measured at the outlet with the help of LDA with LIF capabilities to determine the solid residence times distribution in the FO.

Key Results and Conclusions

Initially, a two-way coupled DPM simulation was performed with 200 µm particles in the FO with an inlet velocity of 2.70m/s (3.5 LPM | τ ~ 1s). The particles were released from the inlet surface, normal to the surface with the same velocity as that of water. Figure 2 shows the velocity contour along with the particle trace and mixing for the different time steps. It was observed that the particles, once they enter the oscillation chamber, they follow the path of the oscillating jet. Appropriate high-speed measurements will be performed to validate the simulated solid-liquid flow dynamics and mixing in FO. The effect of the particles sizes (distribution) on the flow dynamics (mixing) will also be investigated. The modelling and experimental technique given in this work will serve as a base for establishing FO as a crystallizer that will be able to handle solids. The approach shown in this work will also be helpful in designing and optimizing new FOs for a variety of process engineering applications.

Acknowledgement

The authors greatly acknowledge SSPC (https://sspc.ie/), SFI Research Centre for Pharmaceuticals, University of Limerick, Grant number 12/RC/2275_p2) for supporting the work. The authors also acknowledge the support provided by Ansys for CFD licenses under the academic partnership program. The authors also wish to acknowledge the Irish Centre for High-End Computing (ICHEC | https://www.ichec.ie/) for the provision of computational facilities and support

References:

1. Woszidlo, R.; Krüger, O., Editorial for the Special Issue on "Fluidic Oscillators—Devices and Applications". MDPI: 2022; Vol. 7, p 91.
2. Khalde, C. M.; Pandit, A. V.; Sangwai, J. S.; Ranade, V. V., Flow, mixing, and heat transfer in fluidic oscillators. *The Canadian Journal of Chemical Engineering* **2019,** *97* (2), 542-559.
3. Madane, K.; Khalde, C.; Pandit, A.; Ranade, V., Flow Physics of Planar Bistable Fluidic Oscillator with Backflow Limbs. *AIChE Journal*, e17621.
4. Pandit, A. V.; Ranade, V. V., Fluidic Oscillator as a Continuous Crystallizer: Feasibility Evaluation. *Industrial andamp; Engineering Chemistry Research* **2020,** *59* (9), 3996-4006.
5. Yu, Y.; Pandit, A. V.; Robertson, P.; Ranade, V. V., Antisolvent Crystallization using a Fluidic Oscillator: Modeling and Validation. *Industrial andamp; Engineering Chemistry Research* **2021,** *60* (34), 12752-12766.

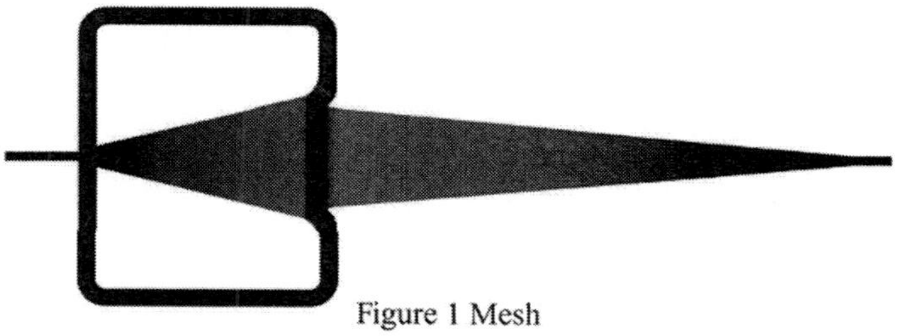

Figure 1 Mesh

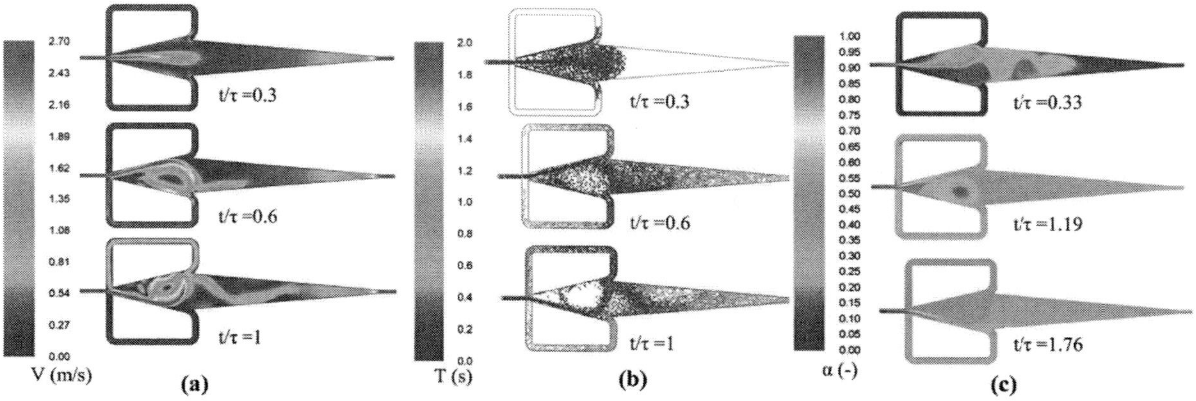

Figure 2 (a) Velocity contours, (b) particle track and (c) mixing

511d Hydrodynamics Characterisation in a Three-Phase Flotation System Using Positron Emission Particle Tracking (PEPT)

Pablo Brito-Parada, Diego Mesa, Katie Cole, Michael R. van Heerden

Impellers play a key role in flotation cells, as the turbulence generated through agitation aids particle suspension, air dispersion and particle–bubble collision. It is therefore important to understand the effect that different impeller designs have on flotation hydrodynamics, as small variations can enhance flotation performance. The study of flotation hydrodynamics is, however, a complex task due to the opaque, multiphase, and polydisperse nature of flotation systems. In this work, the impact of impeller design modifications on the hydrodynamics of a three-phase flotation cell is quantified for the first time. Two different impeller designs, with and without a stator, were assessed using positron emission particle tracking (PEPT), a technique that allows the position and velocity of radioactive particle tracers within an opaque vessel to be determined. A novel PEPT data analysis strategy, supported by a statistical analysis on the basis of the Jensen–Shannon distance, facilitated the comparison of the different designs, allowing a robust quantification of their hydrodynamic differences. The experimental results show that the stator significantly modified the hydrodynamics within the flotation cell, distorting the lower mixing loop that is characteristic of radial impellers. The use of a stator also resulted in the reduction of particle velocity and swirling outside of the impeller–stator region, both at the level of the impeller and, notably, at the pulp–froth interface. These findings have important implications for impeller–stator design, and highlight the effect of the impeller on the hydrodynamics of the pulp and froth phase in flotation systems.

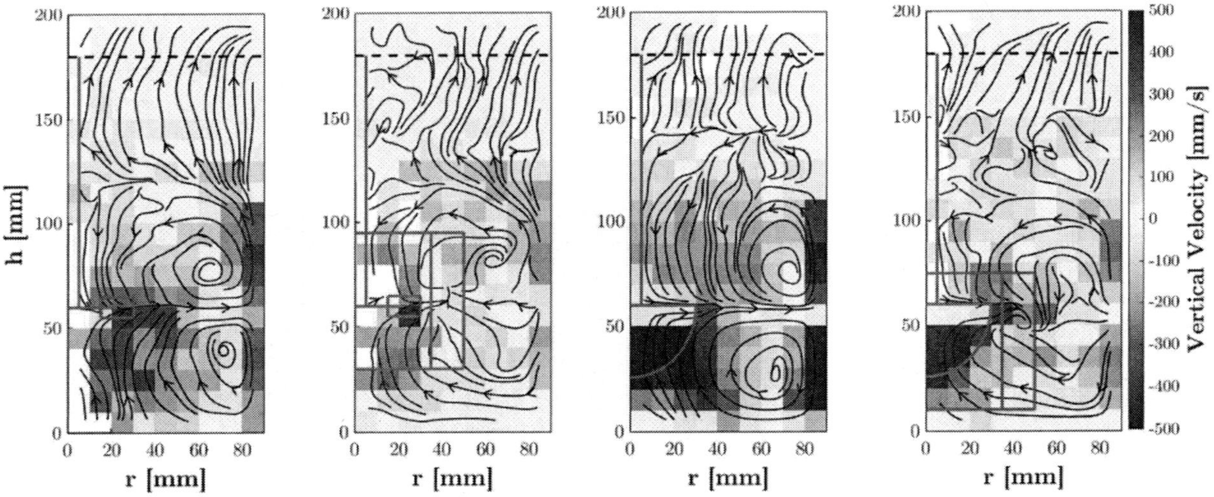

(a) Rushton (b) Rushton-Stator (c) Rotor (d) Rotor-Stator

511e Utilization of Readily Available Computer Programs for Design of Fluid Mixing Equipment

W. Roy Penney

Paper Title:

Utilization of Readily Available Computer Programs for Design of Fluid Mixing Equipment

Author:

1. Roy Penney, Professor Emeritus, University of Arkansas, Department of Chemical Engineering

Abstract:

A design related book was recently published. The reference is Penney, WR "Computer Aided Design of Fluid Mixing Equipment - A Guide and Tool for Practicing Engineers". That 460 page book consists of 15 Chapters that teach the use of 40 Excel Computer Programs, which are used to solve 40 Example Problems, including scale-up problems. That book also presents the technical background to enable practitioners to use 9 computer design programs which currently exist on CalcEdge (access by searching for CalcEdge → Penney); the programs cover: Compartmented Columns, Gas Absorber, Heat Transfer, Impeller Power and Pumping, Liquid-Liquid Dispersions, Pipeline Mixers, Solids Suspension and Vortex Depth. The computer programs used to solve the Example Problems in the Book can be accessed at

https://www.elsevier.com/books-and-journals/book-companion/9780128189757

This paper will really be a paper, which I will distribute widely before the meeting, provided I am allowed access to the Conference Attendance List, with email addresses.

511f Detecting Lagrangian Coherent Structures in Twin-Screw Extruder Elements Via Smoothed Particle Hydrodynamics

Hannes Bauer, Johannes G. Khinast

Introduction

Twin-screw extruders (TSE) are often used to process high viscosity fluids, which are difficult to mix. In the absence of turbulence, the task of creating spatio-temporal disorder (to promote mixing) is achieved with closely intermeshing rotating screw elements. Thus, the geometrical design of these elements is critical for mixing performance. Previous simulation studies have sought to characterize the mixing efficiency via various quantitative measures (i.e., mixing indices). Yet, little is known about the underlying mixing patterns and mechanisms, which hinders improvement of screw designs. Thus, to enhance our understanding, we investigated mixing patterns in TSE elements via Lagrangian Coherent Structures (LCS).

Methodology

Flow simulations were carried out with Smoothed Particle Hydrodynamics[1] (SPH). SPH has proven to perform well for predicting the flow in extruder elements.[2] The method's meshless nature allows for a convenient treatment of rotating moving boundaries such as extruder screws. Additionally, SPH is a Lagrangian method, and as such offers performance advantages for calculating the Finite-Time Lyapunov Exponents (FTLE) fields, which we use for detecting attracting and repelling Lagrangian Coherent Structures (LCS).[3,4]

Results

We simulated the flow in three different TSE elements and determined attracting and repelling LCS for each of them (e.g., Figure 1). The LCS reveal the stretching and folding events that cause efficient laminar mixing. Differences between screw element geometries are discussed.

Literature

1. J. Monaghan, "Smoothed particle hydrodynamics," *Reports Prog. Phys.*, vol. 68, no. 8, pp. 1703–1759, Aug. 2005.
2. Bauer, H., Matić, J., andamp; Khinast, J. (2021). Characteristic parameters and process maps for fully-filled twin-screw extruder elements. *Chemical Engineering Science*, 230, 116202.
3. Haller, G. (2015). Lagrangian coherent structures. *Annual Review of Fluid Mechanics*, 47, 137-162.
4. Sun, P. N., Colagrossi, A., Marrone, S., andamp; Zhang, A. M. (2016). Detection of Lagrangian Coherent Structures in the SPH framework. *Computer Methods in Applied Mechanics and Engineering*, 305, 849-868.

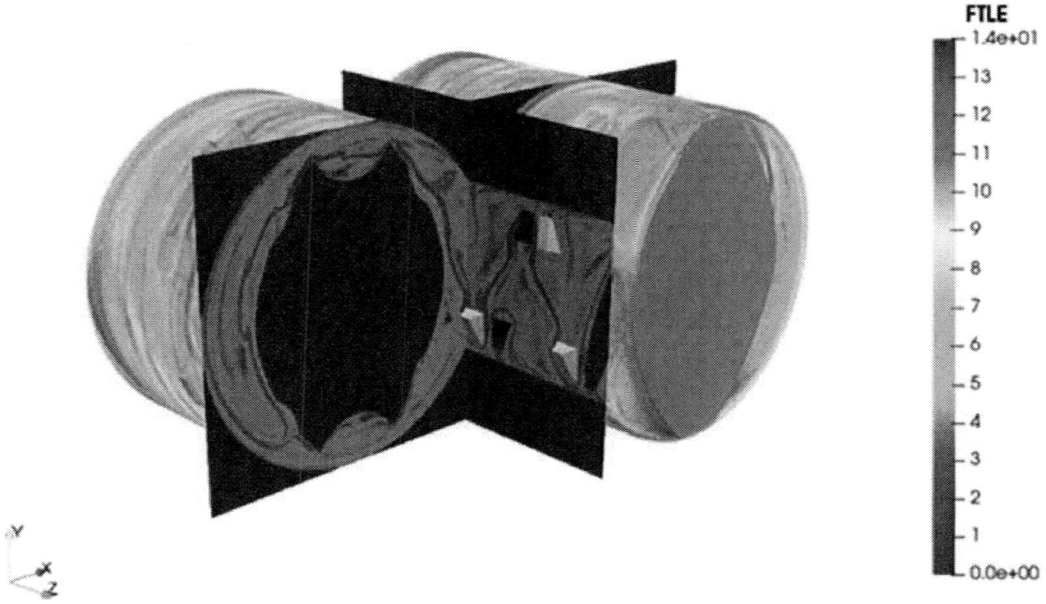

Figure 1: Lagrangian Coherent Structures in TSE elements are detected via Smoothed Particle Hydrodynamics simulations. LCS can be useful in understanding the formation of mixing patterns.

77b A Tribute to Doug Leng - Dow Impact and Legacy

Laura Dietsche, Victor A. Atiemo-Obeng

In this memorial tribute to Doug Leng we first provide an overview of Doug's beginnings at Dow, and highlight the impact he had on the Dow research community. His vision and leadership resulted in the establishment of the Process Fundamentals Research Laboratories at Dow, which have evolved into the Engineering and Process Sciences organization of today, including the Fluid Mechanics and Mixing group. He championed the incorporation of state-of the art experimentalmethods and employing the latest numerical techniques to better understand mixing phenomena in industrial equipment and achieve improve process performance. We highlight his passion for mentoring, training, and developing expertise in the next generation of engineers. His accomplishments, which have been recognized by AIChE through election to Fellow in 1989, by Dow through bestowing on him the prestigious H. H. Dow Medal in 1993, and by NAMF as the NAMF Award winner in 1995, are presented.

77i Douglas E. Leng — An Outstanding Contributor to Dow and to Our Profession

Henry Kohlbrand

Doug Leng was internationally known for his expertise in the field of mixing as it relates to the behavior of chemical processes. This included multi-phase flow (gas-liquid, liquid-liquid and solid-liquid systems), chemical reactors, fluid dynamics, equipment design, problem diagnosis and remedial action, scale-up and related technical topics. Assisted by a strong background in physical fundamentals, he solved a broad variety of technical problems for The Dow Chemical Company on a world-wide basis. Doug was an outstanding mentor and he and his associates made a large economic contribution to Dow and scientific contributions to chemical engineering, He worked for Dow for 40 year(1956–1996). Although known today primarily for his expertise in mixing, over the course of his career he contributed to a number of different technical areas and businesses including, ion exchange products, hollow fiber membranes, renewable resources, chlorinated polyolefins, scale-up, aqueous dispersions, thermodynamic measurements and polymer products. He championed the use of supercomputers in chemical engineering and effectively combined laboratory measurements with fluid mechanics modeling to solve difficult mixing challenges. In 1993 he was awarded the Herbert H. Dow Medal, the highest honor Dow awards to the company's scientists and researchers. Fewer than 30 of these have been given in the history of Dow. This talk will explore Doug's contributions to Dow and his successful career.

77h Six-Year Long Mentorship Under Doug Leng and a Lifelong Friendship

Cassian Lee

This presentation recounts Cassian's experience in working under Doug for six years at Dow's mixing lab. He highlights Doug's kindness and generosity and how Doug had impacted his life. Cassian recollects the state-of-the-art Dow mixing lab and how he got to conduct mixing simulations using CFD under Doug's guidance. Cassian also credits Doug for seeding his passion in videography and photography. Finally, Cassian shares several of his fond memories and stories of Doug, including his last visit with Doug during the Summer of 2019.

77j Memorial to Doug Leng, Mentor and Friend

Allen V. Forster

Many of us know Doug Leng for his long carrier as a PhD, Chemical Engineering Mixing Study Expert, and the many contributions he made to the field. Doug worked for the Dow Chemical Company for over 40 years. But I have known Doug for around 50 years, first working for and with him on a non-Newtonian, Fermentation Mixing Study (in the 70's), intended to increase the production of the Tuberculosis Treatment medication, Rifampicin, produced by a mycelial Streptomycin bacteria. I will talk a little about that study, but I would also like to cover an AICHE event we traveled to and attended together in Chicago, as well as his love of sailing the North Channel of Lake Huron. Over the years, Doug and I also chaired our local AICHE Section. We also preferred using Apple Mac Computers over Widows PC's and chaired those in town clubs as well. The intent of this presentation is not to focus just on Doug's professional career, but on his interests outside of The Dow Chemical Company.

77e Progress on Mixing at Dow Inspired By Doug Leng

David West

Doug Leng was Dow's first mixing expert and an inspiration and mentor for many young researchers interested in fluid dynamics and mixing. When I first met him in the mid-1980s he was a Senior Research Scientist and Laboratory Director of the Central Research Engineering Lab; the only person in the company to have these two high positions on both the technical and managerial ladders. In this talk, I will describe two areas in which Doug inspired and mentored me. I first met Doug about 1987 to seek his advice on a project for expanding Dow's capacity for a raw material (chloroform) used for making so-called soft hydrochlorofluorocarbon refrigerants. The results from research that he had led at the FMP consortium provided a key piece of information needed for designing a reactor capable of increased chloroform production. Some years later he sponsored the development and growth of computational fluid dynamics at Dow. My group benefited from this, especially his support for a unique code written by Bert Harvey which solved the governing equations for laminar flow in a rotating coordinate system. This made it possible to accurately simulate flows in stirred tanks for the first time, accuracy that was established by comparison with detailed velocity-field measurements made in Doug's lab. In collaboration with Doug, we used this capability to design a new impeller which doubled the capacity of one of our polymer plants. Bert and I later developed the first method for computing the mixing time in laminar stirred tanks. In this we converted the computationally intractable problem of solving the time-dependent species transport equation into the problem of iterating a Poincare map; a calculation that could be accomplished in only a few minutes. In this talk, I will describe this problem and solution in some detail.

77c Vision and Influence: Doug Leng's Focus on Computational Fluid Dynamics at the Dow Chemical Company

Joseph Smith

In today's competitive marketplace, business must utilize every tool to make processes more efficient and productive. Although Computational Fluid Dynamics (CFD) has not yet reached routine application in the chemical industry, this tool has great promise. Knowing where and how to apply it is critical. CFD work at The Dow Chemical Company has included polymer processing, crystallization, static/jet mixing, batch mixing, reactor flow optimization, and incineration. This paper presents several examples how CFD was used to solve "real" challenges at the Dow Chemical Company. This work was first presented at the World User's Association for Computational Fluid Dynamics (WUACFD) in Basel, Switzerland in 1993 and 1996. My participation in the WUACFD was made possible by Dr. Doug Leng – a senior research scientist at the Dow Chemical Company in the 1990s. The following projects will be reviewed and discussed in this short tribute to Dr. Leng (Doug) to demonstrate his vision and influence inside the Dow Chemical Company.

The first study examines heated air flow through a "Continuous Through-circulation Dryer" used to produce a super-absorbent used to produce diapers. This work focused on the upward directed flow in the dryer where four fans push heated air up through a permeable bed of "wet" material. Results helped explain why the product had a large range in moisture content. Results also yielded insight into possible ways to improve flow uniformity and increase dryer efficiency and plant capacity.

The second study considered flow into the top of a packed catalytic reactor. This work focused on flow distribution inside the tower head space and its impact on separation efficiency. Uniform flow to the bed surface plus is desired to maximize reactor performance. CFD results illustrated limitations in the current design. The model was used to optimize inlet reactant distribution and to investigate design changes for various inlet flow conditions. This work resulted in improved product yield and reactor throughput.

The last study presented results from our analysis of thermal oxidation of chlorinated hydrocarbons. This analysis utilized CFD with turbulent reacting flow submodels to simulate HCl/Cl2 generation inside the thermal oxidizer. Results of this work resulted in improved system operation with less environmental impact.

These three studies were originally conducted in Midland, MI for the Michigan Division of the Dow Chemical Company. This work was possible because of the vision Dr. Leng had for using CFD to solve real problems for the company.

77d Emulsification and Dispersion in Agitated Liquid-Liquid Systems: Linking Hydrodynamics to Drop Size

Richard Calabrese

Emulsions and dispersions are ubiquitous in the chemical process industries and form the basis of many food, pharmaceutical, personal care and other consumer products. An important step in equipment selection is prediction of drop size for a variety of devices ranging from stirred tanks and static mixers to high shear rotor-stator mills and high-pressure homogenizers. While these devices are of disparate geometry, their analysis can be put on common ground by connecting drop behaviour to local hydrodynamics. Mechanistic models for drop break up and particle fracture are based on the local, scale dependent fluid stress. As a result, having accurate estimates of the extent and intensity of the dispersion zone allows tuning of models for drop size, scaling of drop size data from one device to another of different geometry but with similar local hydrodynamics, and development of physical models for dispersion in the absence of size data.

Scaling arguments will be presented for laminar flow where local shear rate governs drop behaviour, and for turbulent flow where the forces acting on a drop transitions from inertial to viscous as the drop size decreases and local turbulent energy dissipation rate (power per mass) governs drop dynamics. The role of surfactants will be considered. Mechanistic correlations for drop size distribution will be presented and related to local hydrodynamics. The efficacy of different approaches to estimate local deformation rates will be discussed. Detailed flow field results, both experimental and via CFD, will be presented and related to drop or particle dispersion. Population Balance Models (PBM) will be considered with respect to development of breakage kernels and validation of modern CFD-PBM simulations. Examples will be taken from the author's work as well as that of others. Limitations and future directions will be discussed. The presenter was Dr. Leng's co-author on the NAMF Handbook chapter titled "Immiscible Liquid-Liquid Systems". There will be recurring reference to Doug's contributions as well as his influence on the current practice and research directions.

459f The challenges of detailed CFDsimulations ofstirred lab-scalebioreactors

Roya Jamshidian, Harry Van den Akker

Reliable detailed CFD simulations of stirred lab-scale bioreactors are relevant (a) with the view of finding power numbers, as measuring torques at small scales is an issue, (b) because cell cultures usually start in mL vessels, or (c) in the context of designing down-scaled versions of larger-scale bioprocesses. These lab-scale vessels are usually provided with specific impeller types, a single baffle, an aeration device and one or more dip pipes. As a result, a vortex may be formed that interferes with baffle and dip pipes and which turns the flow chaotic, i.e. transient.

In addition, in RANS simulations, the domain is usually divided into two zones: an inner one that 'rotates' with the revolving impeller and an outer stationary zone. To this end, we have the disposal of two options: the steady-state Multiple Reference Frames (MRF) model and the transient Sliding Mesh (SM) approach. Defining such two zones in a small vessel with inserts is a challenge. The presentation will report on our findings when using MRF and SM and testing various grid types and grid densities, with a focus on gas hold-up, energy dissipation rates and spurious velocities.

459b Using CFD to Predict Mixing Coefficients for Scale up

Darren Cappelli

Scaling a process or operation is an important concept in industry. Operating conditions such as agitation rate and gas sparging rate do not produce the same output at different scales. Mixing profiles can vary depending on the scale, which can have a significant impact on process phenomena such as heat, mass transfer, crystallization, and reaction kinetics. Understanding mixing is therefore imperative to understand the process. Mixing can be characterized on three different scales, macro, meso, and micromixing, which correspond to the bulk liquid, droplet, and molecular scales, respectively. The equations describing mixing at these scales can be predicted from mixing time engineering correlations and specific scale-up parameters.

Computational fluid dynamics (CFD) is a useful tool that can be used to provide process insight that experimental analysis cannot. The degree of detail obtained includes velocity distributions, energy dissipation, and macro mixing times which can then be used to predict methods of enhancing mixing efficiency. Vessel-specific mixing coefficients can be regressed for macro, meso, and micro mixing times using the velocity and energy dissipation data from the simulations. Mixing timescales can therefore be predicted for different operating conditions whilst reducing physical experimentation.

Literature values for the mixing coefficients exist however they are for specific reactor-impeller configurations. To date, there are a lack of studies conducted to generate a list of these coefficients and how they vary based on the different reactor variables such as vessel size, impeller type, and the number of baffles. The researcher is therefore forced to choose a coefficient generated in laboratory studies that most closely resemble their system. This introduces inherent error into the calculation and can drastically reduce the accuracy of any mixing time correlation output. For example, the meso-mixing time correlation requires a coefficient that relates the local turbulent eddy dissipation to the power per unit mass. The use of CFD provides an exact measurement of the required coefficient which can be obtained for the bespoke reactor or process under investigation. This improves accuracy and can lead to less experimentation, improved process understanding, and better scale up.

459c A Design of Experiment (DOE)-Based Approach for Bioreactor Scale up and Cell Metabolic Reactions

Sravan Nallamothu, Hossam Metwally, Sivasubramani Krishnaswamy, Sophie Collin, Clementine Shao, Marc Horner

Abstract:

Bioreactors are at the heart of biopharmaceutical manufacturing, and they need to be accurately characterized for optimal scaleup. The transition from lab scale to pilot or production scale with the goal of maintaining the same product yield and homogeneity brings lot of challenges. To evaluate the issues related to scaleup in cell culture bioreactors, one must consider mixing time, oxygen mass transfer rates and shear-stress levels among other factor s. As the reactor scale increases, designers would like to ensure consistency in oxygen mass transfer rate as well as exposure of cells to shear because of their influence on the physiology of the cells.

Optimization-based scale up

In this study, computational fluid dynamics (CFD) simulations are used to predict the mass transfer coefficient ($k_L a$) and shear rates for a lab-scale bioreactor. A response-surface (RS) based optimization algorithm is used to optimize the operation of the bioreactor to maximize the oxygen mass transfer rate while maintaining cell exposure to shear under a specified threshold.

The optimized lab-scale bioreactor is then scaled up to pilot-scale. Similarly, a DOE is set up to scan the design space of the pilot-scale bioreactor. Similarly, a DOE will be used to create a response-surface (RS) which is used to optimize the bioreactor operation, maximize oxygen mass-transfer, and minimize shear exposure.

Cell Metabolism System Model

For both scales, the CFD DOE simulations used to construct the response surfaces will also be used to create reduced-order models (ROM's). The ROM's will take the bioreactor operating conditions as inputs (e.g. rpm and sparge rate) and providing the mass transfer coefficient ($k_L a$) as an output.

Moreover, a 0D transient biological cell growth model was developed that predicts the cell growth, product formation, and gas consumption using the mass transfer coefficient ($k_L a$) predicted from the CFD-based ROM. The 0D transient cell growth model is then combined with the CFD-based ROM's in a system model that predicts cell growth, product formation, and gas consumption vs. Time as a function of bioreactor conditions.

The modeling approach proposed herein is an alternative to scale-up laws which may not always be valid (for example when different impeller types are used at different scales). Additionally, the same DOE used for optimization is used to construct ROM's that are used for system modelling which enables the co-simulation of the flow physics and cell biology.

459d Scale-Down of Viral Inactivation Step: Principles and Practical Challenges

Zbynek Kalal

The low-pH viral inactivation (VI) step is a process used to inactivate potential enveloped viruses in biological drugs manufacturing. The step typically involves a low-pH titration of a chromatography pool using an acid titrant. Defining agitation conditions for the process is important to avoid pH gradients in the protein pool, which could cause higher levels of impurity formation. The typical approach to optimize a low-pH VI step is through utilization of a small-scale model. Developing such a model is complicated by the fact that the flow regime in the small-scale tank may not be representative of the manufacturing-scale system. In this presentation we identify the critical process parameter that defines the process outcome, outline the principles of scaling down the VI process in order to ensure a representative mixing profile in the laboratory system, and highlight the challenge of scaling down to tank sizes that are manageable from a protein utilization perspective.

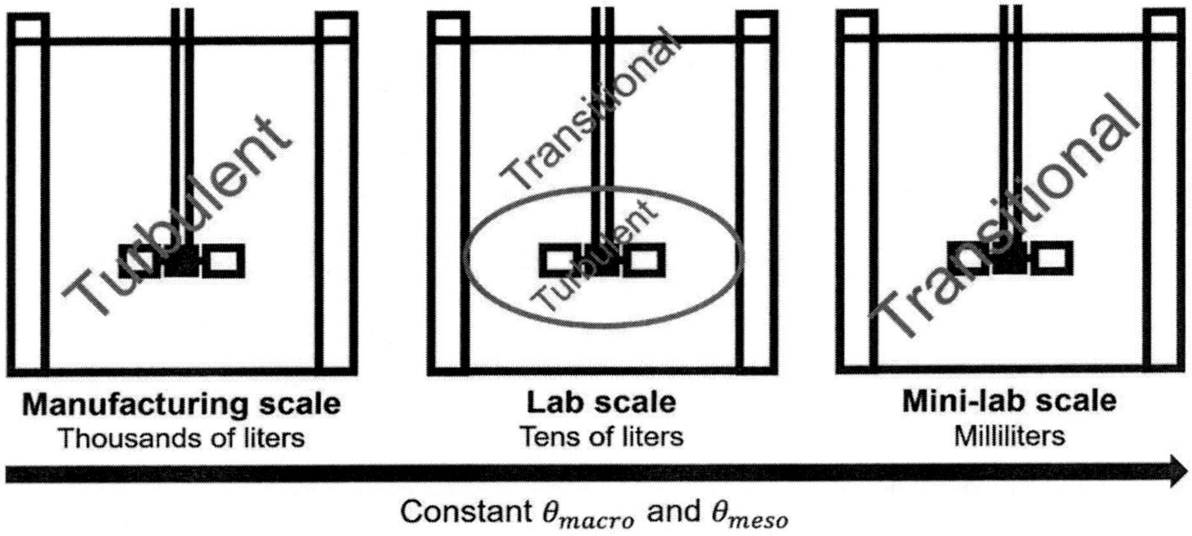

459e Application of Gas-Liquid Mass Transfer Coefficient Fitting Techniques in Batch Reactors for Scale-up

Kira Ragazzo, Jeffrey A. Nye, Shulin Wu, Xiaowen Zhao, Candice Joe, Megan Emmanuel, Antonio C. Ferretti, Richard Heid

Scale-up of hydrogenation reactions can introduce mass transfer limitation, which can impact purity and yield in pharmaceutical scale-up. Knowledge of the gas-liquid mass transfer coefficient (k_La) and mass-transfer limit regime are necessary for reaction rate control during scale-up. However, correlations in literature for k_La are often inaccurate at predicting across various geometries and operating conditions. Therefore, experimental fitting for k_La is often employed. This presentation compares k_La fitting techniques across various geometries during a 1000x scale-up of a nitrobenzene to aniline reduction. Comparison of the techniques is made using kinetic reaction performance. Applicability of each technique is examined with respect to scale, reactor geometry, operating conditions, and analytical capabilities. This analysis enabled appropriate k_La fitting techniques to be employed for predictable and reproducible kinetics during scale-up from 3 g to 3 kg.

AUTHOR INDEX

AIChE
120 Wall Street, FL 23
New York, NY 10005-4020

ISBN 978-1-7138-7893-3
9 781713 878933

CURRAN ASSOCIATES INC.
proceedings.com